I0786934

AI Data Centers and Water Sustainability:

Environmental Impacts, Governance, and Innovation

Hitesh Mohapatra
https://orcid.org/0000-0001-8100-4860

School of Computer Engineering, Kalinga Institute of Industrial Technology (Deemed to be University), Bhubaneswar, India

Amiya Kumar Rath
Biju Patnaik University of Technology (BPUT), Rourkela, India & Veer Surendra Sai University of Technology (VSSUT), Burla, India

Fadi Al-Turjman
https://orcid.org/0000-0001-6375-4123

AI and Software Engineering Department, AI and IoT Research Center, AI and Informatics Faculty, Near East University, Mersin, Turkey

Manjur Kolhar
Department of Health Information Management and Technology, College of Applied Medical Sciences, King Faisal University, Al-Ahsa, Saudi Arabia

IGI Global
Scientific Publishing
Publishing Tomorrow's Research Today

Vice President of Editorial	Melissa Wagner
Director of Acquisitions	Mikaela Felty
Director of Book Development	Jocelynn Hessler
Production Manager	Mike Brehm
Cover Design	Jose Rosado

Published in the United States of America by
IGI Global Scientific Publishing
701 East Chocolate Avenue
Hershey, PA, 17033, USA
Tel: 717-533-8845 | Fax: 717-533-7115
Website: https://www.igi-global.com E-mail: cust@igi-global.com

Library of Congress Cataloging-in-Publication Data

LCCN: 2026031973 (CIP Data Pending)
ISBN13: 979-8-2600-2761-5
Isbn13Softcover: 979-8-2600-2762-2
EISBN13: 979-8-2600-2763-9
British Cataloguing in Publication Data
A Cataloguing in Publication record for this book is available from the British Library.

Table of Contents

Detailed Table of Contents

Due to global demand for advanced technology that requires high-performance computational infrastructure, there has been rapid growth of artificial intelligence (AI) usage. Among this infrastructure are AI datacenters, which consume a large amount of electricity and water. As AI models grow more resource-intensive, the need for water power generation, chip manufacturing, and cooling has risen as critical costs that are overlooked. The paper adopts a critical review of existing literature with a conceptual framework for smart water management in the context of AI-intensive urban infrastructure. The study evaluates how rapid AI development intersects with proposed strategies and freshwater scarcity for reducing water footprints through policy action and technological innovation. The findings highlight an urgent need for transparent water-use reporting, region-specific regulation and sustainable cooling solutions to mitigate the escalating pressure of AI infrastructure on global water resources

The rapid expansion of artificial intelligence (AI) has increased the demand for high-performance data centers, creating major sustainability challenges related to energy consumption, thermal management, and water use. Among these issues,

The explosive worldwide growth of Artificial Intelligence (AI) infrastructures has added new layers to digital infrastructure complexity while generating an unprecedented demand for natural resources especially water. With increasing computational requirements for AI models, data centers need large cooling systems, many of which rely on substantial amounts of freshwater. This chapter investigates its implications for increasing pressures on global water resources. It analyses evolving water consumption trends in relation to regional vulnerabilities, environmental implications and sustainability issues associated with technological growth. The chapter presents a multi policy intervention-level platform approach to balancing AI innovation and responsible water management (through sustainable cooling technology and strategic site selection). By demonstrating the new possibilities for human-centered AI, the results hope to inspire policymakers, technologists, and environmental planners to cultivate more just and sustainable AI-centric futures.

The swift growth of artificial intelligence (AI) infrastructure has introduced important but less explored challenges of water governance. AI water use is growing rapidly and causing an additional strain on other regions & populations that place on water supplies. Through landmark controversies on water use by AI, this chapter examines the global legal controversies on water use by AI and the evolving regulatory responses. The study applies a qualitative methodology based on case law analysis and legal documents to look at how existing legal doctrines like public trust doctrine, beneficial use, precautionary principle, and the right to water are being adapted to such challenges. The focus on environmental justice, corporate power and weak regulation was highlighted. This course will further analyse the way courts and policymakers are creating new legal precedents either to protect technology or the environment. The conclusion of this chapter sets out key legal and policy pathways for responsible and equitable water governance in the AI age.

AI data centers are becoming critical infrastructure, but their growth is also a water-governance issue. Cooling systems can raise local freshwater demand directly, while electricity procurement adds indirect water use across power systems. In water-stressed regions, siting decisions affect not only land and grid capacity, but also basin resilience, transparency, and justice. This chapter develops a governance framework for AI data center growth that combines hydrological screening, public disclosure, and water-justice safeguards. It argues that responsible expansion requires place-based regulation that evaluates cumulative withdrawals, drought sensitivity, emergency curtailment, community participation, and enforceable reporting. Drawing on research on data center sustainability, water disclosure, planning, and environmental justice, the chapter proposes policy instruments for governments, utilities, and operators seeking water-resilient digital infrastructure.

This chapter analyze the environmental, economic, and policy integration of water and energy sustainability in conventional and AI-optimized data centers. The authors seek to study the impacts of architectural decisions and AI optimization on the sustainability performance of data centers focusing on the power usage effectiveness and water-use intensity as the main sustainability metrics and the sustainability trade-offs in the structures capital cost, operational cost, and total cost of ownership over time. The chapter reviews available industry and academic works on the cooling and workload management and driving factors on the increasing water and energy consumption of AI workloads and offers a policy and governance framework to improve the speed of the efficiencies. The responsible use of AI in data centers has the potential to lower operational water and energy consumption. However, AI in data centers has the potential to displace the impacts of sustainability, leading to the justification of the need for the optimization of sustainability processes.

Preface

Artificial Intelligence (AI) has become one of the defining technologies of the twenty-first century, transforming industries, economies, and societies through unprecedented computational capabilities. Behind this digital revolution lies a rapidly expanding network of data centers that support AI model training, deployment, and operation. While much attention has been devoted to the energy demands and carbon footprint of AI infrastructure, an equally critical yet often underexamined issue is its dependence on water resources. Water plays a fundamental role in cooling data center equipment, supporting power generation, and enabling semiconductor manufacturing processes. As AI adoption accelerates worldwide, concerns regarding the sustainability of water consumption associated with AI infrastructure have become increasingly urgent.

The growing intersection between digital transformation and environmental stewardship presents both opportunities and challenges. Data centers are emerging as major consumers of freshwater resources, particularly in regions already facing water stress and climate-related vulnerabilities. At the same time, advances in artificial intelligence, predictive analytics, cooling technologies, and governance frameworks offer promising pathways toward more sustainable resource management. Understanding these complex relationships requires interdisciplinary perspectives that integrate technological innovation, environmental science, public policy, legal studies, and sustainability governance.

AI Data Centers and Water Sustainability: Environmental Impacts, Governance, and Innovation brings together contributions from scholars and practitioners across diverse disciplines and geographic regions to examine the multifaceted relationship between AI infrastructure and water resources. The volume explores the environmental consequences of expanding AI data centers, innovative approaches to water-efficient operations, emerging governance mechanisms, legal and regulatory developments, and strategies for balancing technological advancement with ecological responsibility.

The chapters collectively highlight the urgent need for transparent reporting, sustainable cooling systems, responsible infrastructure planning, water-sensitive policy frameworks, and collaborative governance models. They also emphasize that addressing water sustainability challenges is not solely a technical issue but a societal imperative requiring cooperation among governments, industry leaders, researchers, environmental organizations, and local communities.

CHAPTER OVERVIEW

Chapter 1 presents a predictive analytics–based framework for optimizing water management in AI-driven data center operations. The chapter demonstrates how machine learning and forecasting techniques can improve water-use efficiency, support proactive resource planning, and reduce environmental impacts through data-driven decision-making.

Chapter 2 provides a comprehensive review of the effects of AI data centers on global water resources and introduces a conceptual framework for smart water governance. The chapter examines the growing water demands associated with AI development and proposes policy and technological approaches to reduce water footprints and enhance accountability.

Chapter 3 investigates water consumption patterns and cooling technologies in AI infrastructure. It evaluates conventional and emerging cooling approaches, explores trade-offs between energy and water efficiency, and discusses innovative strategies for reducing the ecological footprint of AI data centers.

Chapter 4 analyzes the freshwater implications of expanding AI data center infrastructure worldwide. Drawing on empirical evidence and existing literature, the chapter assesses regional water-use trends, environmental pressures, and governance challenges while identifying pathways for sustainable water stewardship.

Chapter 5 explores sustainable management strategies for water resources amid the rapid global growth of AI infrastructure. The chapter examines environmental and regional vulnerabilities and proposes policy interventions, technological solutions, and planning approaches that support responsible AI development.

Chapter 6 focuses on legal disputes and regulatory responses related to water use by AI infrastructure. Through the examination of landmark cases and emerging legal doctrines, the chapter highlights evolving precedents, environmental justice concerns, and future directions for equitable water governance.

Chapter 7 develops a governance framework for AI data center expansion in water-stressed regions. Emphasizing transparency, community participation, hydrological assessment, and water justice, the chapter proposes policy instruments that promote responsible and resilient digital infrastructure development.

Chapter 8 examines the interconnected challenges of water and energy sustainability in both traditional and AI-driven data centers. The chapter analyzes environmental, economic, and policy dimensions of sustainability while evaluating how AI optimization can simultaneously improve efficiency and introduce new governance considerations.

Taken together, the chapters in this volume offer a comprehensive examination of one of the most pressing sustainability challenges emerging from the AI era. By addressing technological innovations, environmental impacts, governance frameworks, and legal dimensions, this book contributes to the growing discourse on responsible AI infrastructure development. It is our hope that the insights presented here will inspire further research, inform evidence-based policymaking, and support the creation of digital ecosystems that are both technologically advanced and environmentally sustainable.

As AI continues to reshape the world, ensuring the responsible stewardship of water resources will remain essential to achieving long-term sustainability. This volume seeks to advance that important conversation and provide readers with the knowledge and perspectives necessary to navigate the evolving relationship between artificial intelligence, data centers, and water sustainability.

Chapter 1
Predictive Analytics-Driven Water Management in AI Data Center Operations:
A Sustainable Intelligence Framework for Resource-Constrained Environments

C. V. Suresh Babu
http://orcid.org/0000-0002-8474-2882
Hindustan Institute of Technology and Science, India

S. Lokesh
http://orcid.org/0009-0001-3736-9157
Hindustan Institute of Technology and Science, India

Lakshmi Narayanan
Hindustan Institute of Technology and Science, India

J. Ajay Aravind
Hindustan Institute of Technology and Science, India

ABSTRACT

This chapter per the authors examines the role of predictive analytics in optimizing water management within AI-driven data center operations, addressing the growing environmental concerns associated with water-intensive cooling systems. It is intended for researchers, data center engineers, sustainability planners, and policy

DOI: 10.4018/979-8-2600-2761-5.ch001

stakeholders. The study adopts a conceptual and analytical approach, integrating machine learning–based forecasting models with infrastructure and environmental data to propose a sustainable water management framework. The findings indicate that predictive techniques can significantly improve water-use efficiency, enable proactive resource planning, and reduce environmental impact. The chapter concludes that data-driven water optimization is essential for building sustainable AI infrastructure and informs future policy and operational strategies.

1. INTRODUCTION

The rapid development in artificial intelligence (AI) technologies has also led to a rapid development in data center infrastructure across the world. Data centers are considered the computational core for cloud computing, machine learning, and data processing systems. In order to meet the growing digital demands, data centers have become highly dense and high-performance environments that require significant computational and physical resources to function effectively (Suresh Babu C. V, Sathiya Narayana S. 2024).

What makes this paper different from other works is that the authors suggest the implementation of an analytics-driven predictive model designed for forecasting water consumption in AI-powered data centers, unlike previous studies based on traditional energy-efficient optimization approaches (Mytton, 2021; Lei et al., 2025). The novelty lies in the consideration of factors like workload, environmental impact, and cooling efficiency within a single prediction loop (Zhang et al., 2021; Masanet et al., 2020). The model includes WUE and environmental variables in its algorithmic architecture, allowing it to adapt to changes and sustainably manage infrastructure resources (Jiang et al., 2025).

1.1 Growth of AI Data Centers and Infrastructure Expansion

The rise of AI-based application tools such as deep learning tools, real-time analytics tools, and generative tools has resulted in a significant increase in computational workloads. This has resulted in the development of hyperscale data centers with high processing density and constant operation. These data centers require a significant amount of energy and water resources, with the amount of resources being directly proportional to the computational workloads (Gao et al., 2012; Rongon & Das, 2025). Research has shown that the amount of energy consumption in global data centers has been rising steadily over the years, owing to both the rise of digital services and the complexity of AI-based workloads (Masanet et al., 2020; Koomey, 2011, Liu et al., 2020). According to recent estimates of hyperscale data centers, these facilities require a large amount of water resources for operation, with the amount of water resources varying considerably based on climate conditions.

While extensive research exists regarding the sustainability of data centers, most of the existing literature is focused on the efficiency of energy consumption with little attention given to modeling water usage prediction (Masanet et al., 2020; Mytton, 2021). Moreover, present-day models fail to incorporate environmental factors like weather conditions and water stress levels in different regions into their infrastructure planning (Mekonnen & Hoekstra, 2016; UNESCO, 2023).

Although there have been significant developments in the field of AI-based data centers, a pertinent research problem arises when considering the amount of consideration devoted to water consumption as opposed to energy consumption. Current methods mostly focus on how efficient the energy consumption is without taking into account the fluctuating and environmental nature of the consumption of water in the cooling system.

1.2 Water Consumption as an Emerging Sustainability Challenge

Whereas energy efficiency has long been a matter of study and research, the idea of water usage in data centers is something which has only lately been understood to be a matter of concern in sustainable development. Water consumption happens through cooling operations to deal with the heat that is produced from high-performing computing machinery. Cooling processes may use many kinds of techniques, among which are the use of evaporative cooling and liquid cooling systems. Large quantities of water have been consumed during such processes for cooling purposes. Studies conducted have shown that water consumption in data centers depends on many factors including the environmental ones.

The increasing dependence on water resources is of special concern in areas already facing water scarcity issues. As per global reports and assessments, billions of people worldwide are facing severe water stress issues, and this is of special concern regarding the long-term sustainability of water-intensive digital infrastructure (Mekonnen and Hoekstra, 2016; UNESCO, 2023). This is posing a direct challenge to technological advancements and their contribution to environmental protection and conservation. As per reports and studies conducted so far, water consumption is a major concern for technological development; however, these reports and studies are mostly focused on energy optimization and not water predictive modeling (Mytton, 2021; Lei et al., 2025).

The existing literature on the sustainability of data centers can be broadly classified into two categories: energy-centric optimization models, and recently, emerging studies on the analysis of water consumption. Though the former category dominates the literature, recent studies have started to reveal the importance of the latter, yet without the use of predictive analytics for optimization.

1.3 The Intersection of AI Innovation and Environmental Responsibility

In this context, artificial intelligence represents both a problem and an opportunity. On one hand, it represents an added problem in the increasing demand for computation, thus exacerbating the problem of resource consumption. On the other hand, it represents an opportunity through the application of advanced data science approaches to optimize infrastructure. In this regard, predictive analytics represents an opportunity to forecast future demand and identify areas of inefficiency. This is an important opportunity in terms of infrastructure management since it provides an opportunity for predictive analytics to be used for decision-making.

Using machine learning models, data center operators can shift from reactive management strategies to predictive and adaptive management strategies. The strategies can help in the efficient use of water resources, cooling strategies, and environmental responsibility. The use of predictive analysis in data center management is consistent with the overall objective of creating a sustainable environment, which is the essence of developing an environmentally friendly digital ecosystem (Cina et al., 2025).

1.4 Chapter Structure and Overview

This chapter seeks to investigate the use of predictive analytics in the improvement of water management in AI-based data center operations. It starts by discussing the link between the architecture of a data center and its use of water resources; it then discusses the use of predictive analytics techniques and how they are used in managing infrastructure. Finally, it provides a conceptual framework for modeling and optimizing the use of water resources (Islam et al., 2025).

The following sections elaborate on the implications of predictive water management from the point of view of operations, environment, and policies, and offer some recommendations for data center operators and policymakers. Finally, the chapter concludes with some possible avenues of future work and the necessity of sustainability while designing infrastructures for AI systems.

2. AI DATA CENTERS AND WATER DEPENDENCY

The AI-based data centers are advanced infrastructure, designed for the operation of high-performance computing. The efficiency of the infrastructure largely depends on the computational architecture, as well as the thermal management systems. Among the essential resources for the operation of the AI-based data centers is the availability of water, particularly for the purpose of cooling.

2.1 Architecture of AI-Driven Data Centers

The contemporary AI data centers have been designed to support the execution of parallel processing tasks in large quantities. This includes the training of deep learning models and the execution of real-time inferences. This is usually achieved through the integration of high-density servers, which have the support of GPUs. This architecture ensures the integration of various components, which operate in real-time to support the high computational needs of the systems (Dayarathna et al., 2016).

The hyperscale data centers have enhanced the existing design by the integration of thousands of servers in a single facility. This has enhanced the power density and heat dissipation. The main focus of the design is to ensure scalability, redundancy, and efficiency. However, these factors have enhanced the resource consumption of the systems, especially the cooling systems (Masanet et al., 2020).

2.2 High-Performance Computing and Heat Generation

The workloads of AI have high computational requirements and produce considerable heat due to continuous computation and high utilization of machine resources. Graphics processing units and tensor processing units are commonly used in AI applications. They have high power density and produce considerable heat.

The association of computational intensity and heat generation is directly proportional. In other words, the more the workload, the more the requirement for heat dissipation. In the absence of such heat dissipation, the heat generated by the system can adversely impact the performance of the system and reduce the lifespan of the hardware. This can also lead to failures in the system (Gao et al., 2012). In such a case, the importance of thermal management in the operation of the data centers is evident.

Industry reports have suggested that the hyperscale data centers of the major cloud providers have experienced high increases in the requirement for cooling. This is attributed to the operation of AI workloads.

2.3 Cooling Technologies and Water-Intensive Mechanisms

Cooling systems are critical for maintaining optimal operating temperatures in data centers. The cooling system is different in terms of efficiency, but they are generally water-cooled.

2.3.1 Evaporative Cooling Systems

The most commonly used technique in the data centers is evaporative cooling. The evaporative cooling technique is the process through which the evaporation of water occurs, which helps in the reduction of the ambient temperature. This is an energy-efficient technique; however, the amount of water required is more, especially if the data center is larger in size and the climatic conditions are hot (Zhang et al., 2021).

2.3.2 Liquid Cooling Technologies

Liquid cooling systems use the direct transfer of heat from electronic equipment through liquid mediums such as water and other coolants. It is a more effective cooling system than air cooling since liquids have higher thermal conductivity than air. It also creates a need for water resources and sophisticated infrastructure for liquid circulation and heat transfer (Zhang et al., 2021).

2.3.3 Hybrid Cooling Approaches

The hybrid technology involves a combination of air and water. The use of these technologies is aimed at attaining the best results under different conditions. In spite of the fact that the use of this technology minimizes the use of water, there is still a necessity for water in the process. New technologies such as immersion technology and advanced air technology have been proposed to eliminate the use of water. However, the use of these technologies is still in its infancy due to cost constraints.

Table 1. Comparative Analysis of Cooling Technologies in AI Data Centers

Cooling Technology	Water Consumption Level	Energy Efficiency	Operational Complexity	Suitability for AI Workloads	Sustainability Impact
Air Cooling	Low	Moderate	Low	Limited for hyperscale AI workloads	Moderate
Evaporative Cooling	High	High	Moderate	Suitable for large-scale AI operations	Lower due to heavy water usage
Liquid Cooling	Moderate	Very High	High	Highly suitable for GPU-intensive workloads	High
Hybrid Cooling	Moderate	High	Moderate to High	Flexible for dynamic workloads	Improved balance of efficiency and sustainability
Immersion Cooling	Very Low	Very High	High	Emerging solution for advanced AI systems	Very High

2.4 Water Usage Effectiveness (WUE) and Performance Metrics

Water Usage Effectiveness (WUE) is another key factor used in the assessment of water efficiency in data centers. WUE is defined as the ratio of water used to the amount of energy used (Jiang et al., 2025). Thus, it can be noted that the lower the WUE, the better it is. WUE provides a way of measuring water efficiency in data centers in an organized manner. This provides an opportunity for benchmarking. However, it is noted that WUE does not take into account regional water availability and environmental concerns. This reduces its effectiveness (Suresh Babu, C. V., Subha Sree, et. al., 2024)

Furthermore, the relationship between the consumption of water and the consumption of energy in data centers can be considered interdependent, where the increased consumption of water for the purpose of cooling will lead to increased consumption of energy, thus establishing the direct relationship between the two in the context of environmental impact, i.e., carbon footprint. This interdependent relationship, therefore, establishes the indirect impact on carbon footprint due to increased water efficiency, thus the need for integrated optimization. Although WUE is a useful metric for benchmarking water efficiency, it does not fully capture regional water scarcity, environmental impact, or dynamic operational variations, thereby limiting its effectiveness as a standalone sustainability indicator (Jiang et al., 2025)

2.5 Regional and Global Water Stress Considerations

The environmental impact of data center water use is also highly dependent on the geographical location of the data center. Data centers located in water-scarce areas are likely to increase the problems of water scarcity.

According to studies conducted globally on data centers, a large percentage of the world's population is facing water scarcity problems. Therefore, data centers are becoming unsustainable in certain areas of the world (Suresh Babu, C. V., Subha Sree, et. al., 2024) In addition, there is a lack of standardized reporting and regulation of data center water use (UNESCO, 2023).

Figure 1. Relationship Between AI Workloads, Heat Generation, Cooling Systems, and Water Consumption

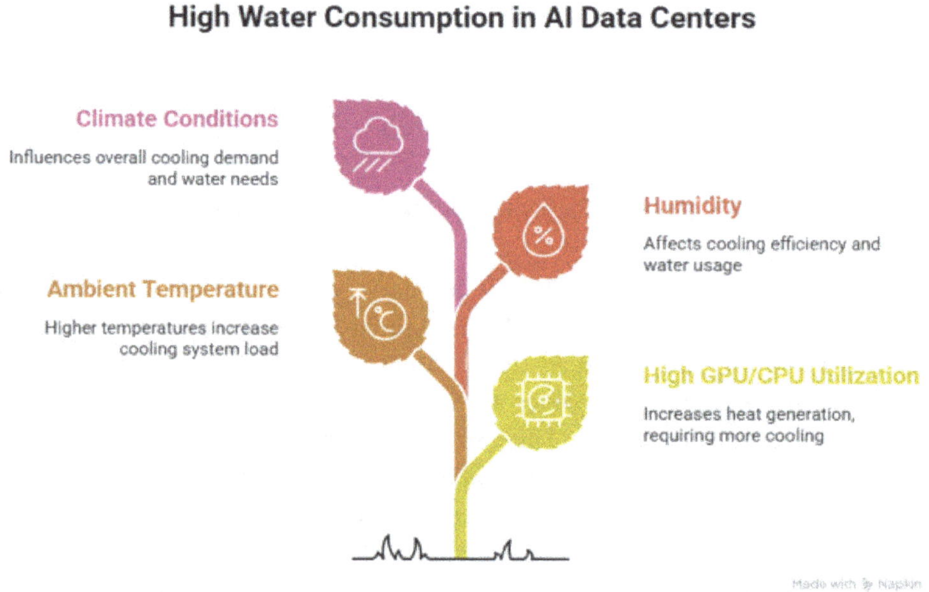

3. PREDICTIVE ANALYTICS: CONCEPTS AND FOUNDATIONS

Predictive analytics has been identified as a significant tool in the management of infrastructure, as it promises data-driven decision-making using data analysis, which involves the analysis of historical and real-time data (Kurniawan & Anouzla, 2026).

3.1 Predictive Analytics for Infrastructures Management

Predictive analytics is defined as "the use of statistical models, machine learning, and data mining techniques to make predictions by identifying patterns or trends in the analysis of data." The role of predictive analytics in managing infrastructures is that the technique helps to ensure efficient operations of infrastructures because of the predictive or identifying nature of trends, demand, or patterns within infrastructures (Cerquitelli et al., 2020; Sarker, 2021).

In the management of data centers, predictive analytics has traditionally been used for the optimization of the usage of power or the management of the workload. However, the use of predictive analytics for the management of water resources has not been sufficiently explored. In this regard, it is possible to assert that the relationship between the computational workload and the heat output or the cooling requirement is a basis on which the predictive model could be extended for the prediction of the water usage.

However, the majority of the predictive analytics implementations in data centers have focused on the optimization of power usage, with a lack of emphasis on the prediction of the usage of water. This is a basis on which it could be asserted that there is a lack of development of a domain-specific predictive model (Suresh Babu, C. V., Subha Sree, et al., 2024).

In the real-world implementation of the predictive model, the accuracy of the predictive model is highly dependent on the quality of the data available for the predictive model. The predictive model is developed based on the available data sets, which are then updated periodically using real-time data for the purpose of adaptive forecasting.

3.2 Machine Learning Techniques for Forecasting Resource Consumption

Regression and time-series models are particularly suitable for modeling resource consumption due to their ability to capture continuous trends and temporal dependencies in operational data (Leon et al., 2020). The application of machine learning is also beneficial in data-driven forecasting for the consumption of resources in data centers, especially in water demand modeling.

3.2.1 Regression Models

Regression-based methods can be used for prediction of any type of continuous data, such as water consumption. Linear regression and its extensions can be used for modeling the relationship between the features such as workload, temperature, and cooling system. It is computationally efficient and easy to interpret for water consumption forecasting, as explained in the article by (Suresh Babu, C. V., Yadav, S., et. al. 2025).

3.2.2 Time-Series Forecasting

The time series techniques are useful for dealing with dependencies in data. ARIMA and RNN, and their variations like LSTM, are effective models for dealing with dependencies in data, like water consumption. These models are effective in accurately predicting the water consumption by identifying trends and peaks, as mentioned in the article by Leon et al. (2020).

3.2.3 Anomaly Detection Methods

Anomaly detection techniques are used for the identification of anomalies or irregularities in the use of water. Anomaly detection techniques play an essential role in the identification of any leaks or inefficiency in the use of water. Clustering, isolation forest, and neural network-based anomaly detector machine learning techniques can be used for the early identification of anomalies or irregularities in the use of water and to take corrective action (Sarker, 2021). Even though the regression model is essential for its accuracy and simplicity, the use of time series and deep learning-based models is essential for their accuracy in the identification of variations in the use of water (Suresh Babu, C. V., Yadav, S., et. al. 2025).

3.3 Role of Real-Time Data and IoT Sensors

Predictive analytics will be effective only when there is quality data available. Nowadays, in the case of any data center, sensors become crucial for obtaining timely data about temperature, humidity, workloads, and even water consumption using Internet of Things (IoT) sensors. These sensors make sure that data regarding such parameters is obtained continuously to ensure that there is quality data available for predictive modeling (Shah et al., 2021).

Real-time data integration helps in the dynamic adaptation of the predictive models according to the changing conditions of the environment. For instance, any change in the external temperature or workload can be factored in while building the predictive models for more efficient use of resources such as water and for the better sustainability of the environment.

3.4 From Reactive Monitoring to Proactive Intelligence

However, unlike the reactive nature of the traditional approaches, predictive analytics makes it possible to move towards proactive intelligence, where potential problems can be anticipated and solved in advance. This is possible by joining machine learning models with live data feeds. This shift not only eliminates the wastage of resources, potential operation risks, and the improvement of the system's overall robustness, but also makes it possible to implement more advanced frameworks such as the hybrid learning model and federated learning approaches, which make the most of the distributed nature of the data while ensuring the privacy of the information (Qi et al., 2023; El Hanjri et al., 2023).

4. MODELING WATER CONSUMPTION IN AI DATA CENTERS

However, to formulate a model for water consumption in AI data centers, it is essential to have a wide-ranging knowledge of factors that influence water demand. This is as opposed to a static system where there is no dynamic process for water consumption. This is due to the complex relationship between computational workloads and environmental factors. A predictive approach to modeling is a method for building a correct model of water demand. There are a number of factors that influence water demand in data centers. It is essential to identify these factors to ensure a correct model.

4.1 Determinants of Water Usage Patterns

These variables can be treated as input features in predictive models, enabling the formulation of data-driven relationships between system conditions and water consumption.

4.1.1 Workload Intensity and Computational Density

The workload that depends on AI technologies like deep learning and processing data at a massive scale produces heat because of the high usage of computer systems resources. An increase in the amount of workload in intensity results in an equal increase in cooling requirements. Thus, there will be an increased water requirement for cooling (Masanet et al., 2020; Lei et al., 2025). The density of computing depends on the density of computing units per unit area.

4.1.2 Climate and Environmental Variables

The external environment, such as temperature, humidity, and seasonal variations, is also important when considering the need for cooling. Data centers located in a warm climate have a higher need for cooling; therefore, the water consumption rate is also higher for such locations. In addition, the level of humidity affects the need for evaporative cooling; therefore, it affects the level of water consumption.

4.1.3 Factors Influencing the Design of the Infrastructure

Another factor that affects the usage of water is the design of the infrastructure itself. In this regard, the cooling approach for the data center may involve air cooling, liquid cooling, or a hybrid of both methods. In addition, the rate of heat dissipation is another critical determinant of water usage (Zhang et al., 2021).

For example, changes in the intensity of workload and temperature can cause large variations in water demand. In addition, peak computational workload may coincide with high temperatures.

For instance, peak computational workload and high ambient temperature can cause large variations in water demand. In this case, there is a potential for increased water consumption during peak operational times.

4.2 Sources of Data for Predictive Modeling

Predictive modeling is an effective tool that requires various sources of data to be used efficiently. Some of the data sources include:

- **Operating data** – server performance, workload, and power consumption
- **Environmental data** – ambient temperature, humidity, and weather conditions
- **Infrastructure data** – cooling efficiency, water velocity, and system design

IoT-enabled sensors will play an essential part in gathering real-time information in these sectors. Real-time data is essential for the updating of the models. Real-time and historical datasets have been widely used for the training of machine learning-based models. This concept of using various forms of data for the management of resources in various systems has been proven in the management of large-scale industries and infrastructures (Suresh Babu C. V, Sathiya Narayana S. 2024, Cerquitelli et al., 2020).

Table 2. Determinants of Water Consumption Patterns in AI Data Centers

Variable	Explanation of the Variable	Impact on Water Consumption
Intensity of Work	Degree of computational intensity	Higher work intensity generates more heat, increasing the need for water-based cooling
Temperature	Ambient or operating temperature	Higher temperatures require more cooling, leading to increased water usage
Humidity	Atmospheric moisture level	Affects the efficiency of evaporative cooling systems
Efficiency of Cooling	Effectiveness of the cooling system	Higher efficiency reduces water consumption
Type of Cooling	Cooling methodology used	Determines the baseline water requirement (e.g., air cooling vs. liquid cooling)

4.3 Framework for Forecasting Cooling-Water Demand

The predictive framework for water consumption forecasting usually comprises the following three major stages:

Data acquisition, model development, and decision support.

During the data acquisition phase, the data is collected from the various sensors and monitoring systems. This phase also includes the preprocessing of the collected raw data to reduce the noise and inconsistencies.

The next step will be the model development phase. In this phase, the development of different machine learning methods such as regression analysis, time series analysis, etc., will be carried out. It has been proposed that advanced machine learning methods such as ensemble learning and deep learning will be utilized for the improvement of the precision of the forecasting for the water consumption, taking into consideration the complex relationships between the factors such as water consumption and input factors such as workload and temperature (Suresh Babu, C. V., Yadav, S., et. al. 2025).

The decision support stage will be the stage where the predictions will be incorporated into the workflow of the operation. The predictions for the demand for the water will be utilized for the optimization of the cooling process so that the resources will not be overused.

In a simulated environment, the predictions can be evaluated by comparing forecasted water usage with actual consumption across different workload scenarios using standard validation approaches.

An abstract expression for predicting water usage could be formulated as:

WU = f (Workload Intensity, Temperature, Humidity, Cooling Efficiency)

This would mean that the model could take a regression format as follows:

$$WU = \alpha W + \beta T + \gamma H + \delta C + \varepsilon$$

In this model, W stands for workload intensity, T is the environmental temperature, H is humidity, and C is the efficiency of the cooling system, while α, β, γ, and δ are coefficients obtained through regression analysis. Such an approach fits within the framework of regression-based predictions for resource consumption (Suresh Babu, C. V., Yadav, S., et al., 2025; Leon et al., 2020).

Figure 2. Predictive Analytics Pipeline for Cooling-Water Demand Forecasting

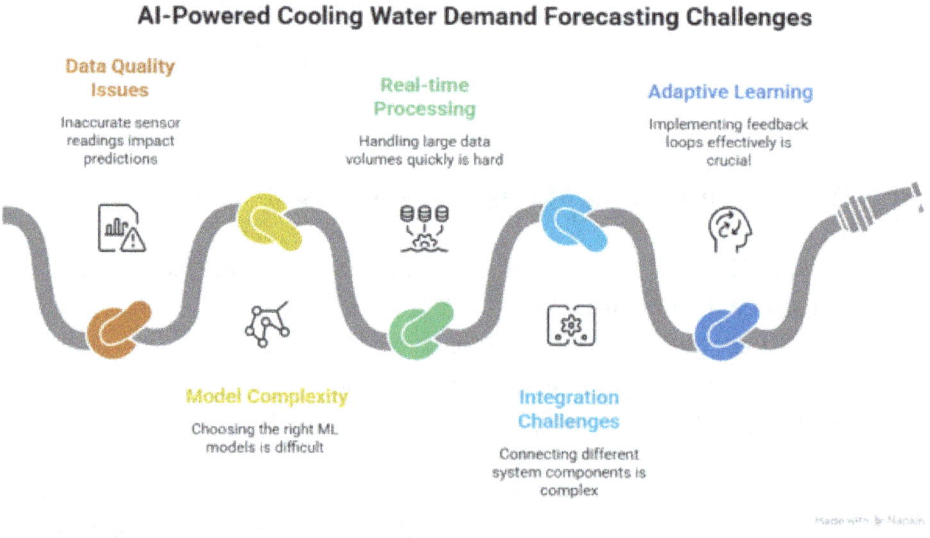

4.4 Integrating Climate and Hydrological Data into Predictive Systems

In order to enhance the accuracy of the models, it is important to incorporate the information related to climate change and hydrology into the models. The information related to climate change includes the availability of water in different regions of the world, drought conditions, and climate change over a period of time. All these pieces of information will help the data center operators make decisions that are compatible with the environment.

In addition, the models will be able to adjust the cooling mechanism of the data center according to the scarcity of water or environmental conditions, thereby reducing the impact of the data center on the environment (Kollar & Grady, 2025).

Such models will ensure the sustainability of the infrastructure and will comply with environmental laws and policies (UNESCO, 2023; Mekonnen & Hoekstra, 2016).

In brief, the modeling of water consumption in AI data centers involves a multi-dimensional approach. The predictive analytics tools help in the accurate forecasting of water consumption. The environmental, technical, and operational factors need to be taken into account while using these tools. Scenario-based modeling can help in the decision-making process by analyzing the best and worst case scenarios of water consumption. The potential efficiency improvements can be estimated.

4.5 Methodological Approach

For the analysis of water consumption at AI data centers, a modeling and simulation method will be applied. Specifically, the proposed modeling approach will be developed based on the conceptual modeling and empirical assumptions derived from previous works concerning water consumption in data centers.

The inputs which may influence the water consumption include:

- Workload intensity (CPU/GPU utilization)
- Environment parameters (temperature and humidity)
- Parameters related to cooling systems

The regression-based model will be constructed as an approach. It is important to emphasize that there will not be a specific training set for this model. Specifically, the relationships found during the course of the previous researches (Leon et al., 2020; Suresh Babu, C. V., Yadav, S., et al., 2025) will be used to build it.

Apart from the abovementioned approach, the hypothetical case of simulation will be constructed in order to demonstrate the change of water consumption due to variations in the workload intensity and environment parameters. For these purposes, only the directionality of changes will be used.

5. PROPOSED FRAMEWORK FOR PREDICTIVE ANALYTICS FOR SUSTAINABLE WATER MANAGEMENT

The framework discussed in this paper is more conceptual in nature. It is a combination of architecture and mathematics. The paper makes no use of a particular set of data from the real world but gives a general framework for making predictions using predictive analytics that can later be implemented in AI-powered data centers. As a solution to the problem of water consumption in AI data centers, a framework has been devised that focuses on a predictive analytics system. The combination of predictive analytics systems, machine learning algorithms, real-time systems, and decision support systems would guarantee efficient water management.

5.1 Conceptual Framework Design

The proposed framework is based on a closed-loop framework, which includes data acquisition, predictions, and strategy adaptation. The proposed framework is based on the following three levels:

The first is the acquisition of data; data is acquired through various means such as IoT devices, environment sensors, and data centers' operation. This includes various factors such as the intensity of the workload, temperature, humidity, and water flow rates.

The predictive intelligence layer makes use of machine learning algorithms for the prediction of the demand for water. It helps in identifying the inefficiencies within the system. It does this by analyzing the data, both historical and real-time, and identifying the trends for the coming days (Suresh Babu C. V, Sathiya Narayana S. 2024).

The decision optimization layer makes use of the data obtained from the predictive intelligence layer. It optimizes the decisions for the efficient use of water without compromising its performance.

5.2 Architecture of Predictive Water Intelligence System

The architectural framework of the discussed system is composed of multiple layers interacting with each other as follows:

- The first layer of sensors gathers live data from various parameters using IoT technology (Shah et al., 2021).
- The second layer of data processing uses collected data to create attributes used for future analysis within machine learning models.
- The third layer of modeling uses different regression, time-series forecasting, and machine learning models.
- The fourth layer of control regulates the operation of cooling systems based on the data generated by machine learning models.
- The fifth layer of visualization & monitoring makes available information about water consumption.

Figure 3. illustrates the proposed predictive analytics framework for water management in AI data centers. Alternatively, for an improved version of the model, distributed learning techniques can be applied, specifically, federated learning can be employed as a way of optimizing the use of the data collected at facilities without affecting the level of data privacy (El Hanjri et al., 2023). In practice, it can be tested in a real-world setting by employing historical operational data and comparing the predicted data on water consumption with the real data on consumption.

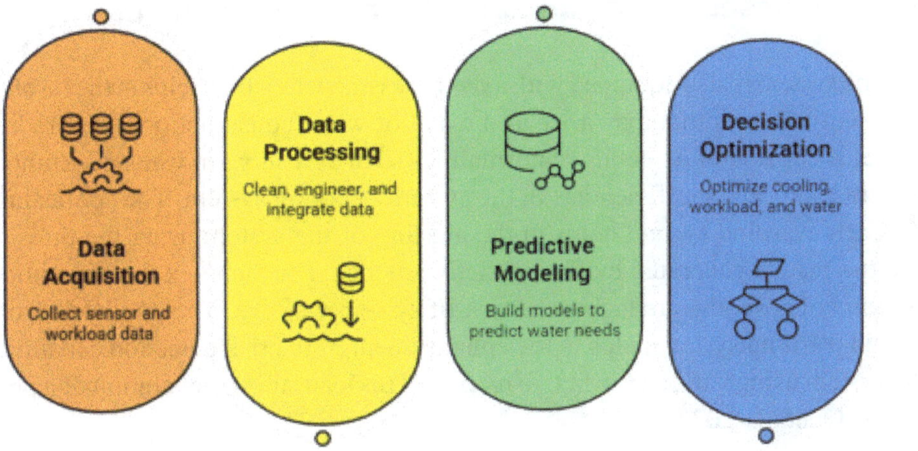

5.3 Key Performance Indicators for Water Optimization

To measure the effectiveness of the framework, some key performance indicators (KPIs) have been identified as follows:

- Water Usage Effectiveness (WUE): This KPI is utilized to evaluate the overall water usage efficiency concerning IT energy usage (Mytton, 2021).
- Prediction Accuracy: This KPI is utilized to evaluate the accuracy of the prediction model concerning some metrics such as MAE and RMSE.
- Cooling Efficiency: This KPI is utilized to assess the effectiveness of the system in ensuring efficient cooling performance while minimizing water usage.
- Anomaly Detection Rate: This KPI is utilized to assess the anomaly detection performance of the system concerning abnormal water consumption, which causes inefficiency.

Thus, from the KPIs, it is quite clear that there is an overall evaluation of effectiveness in terms of sustainability. By using hypothetical scenarios, in a hyperscale data center, predictive models could be used to evaluate the maximum demand for cooling. This would further enhance Water Usage Effectiveness (WUE), which would reduce the amount of water used during non-peak hours. By using hypothetical scenarios, predictive models could be used to obtain quantified results in enhancing water efficiency by minimizing non-essential water usage.

5.4 Risk Assessment and Early Warning Mechanisms

The system is also equipped with a risk assessment tool that helps in the resolution of potential issues that may arise as a result of water consumption. The predictive model is capable of detecting abnormalities such as a surge in water consumption, system leaks, and inefficiencies in the system's cooling system. The application of the early warning system helps in the alerting of the system using the thresholds and the results generated by the predictive model. This helps in the prevention of system collapse, thus reducing the risk of system wastage. The application of the hybrid learning system helps in the enhancement of the risk detection capability of the system using various predictive models. This leads to the generation of a precise result (Qi et al., 2023).

5.5 Alignment with Sustainability and Environmental Governance Goals

The framework, which is proposed, is a development of the various methodologies used in predictive analytics, which include the addition of water forecasting with specific environmental parameters. It is a framework that bridges the gap in the various research conducted on sustainability in data centers. It assists organizations such as data centers in developing strategies on how best to reduce their impact on the environment, especially in areas where there is a water scarcity (UNESCO, 2023). The framework is also unique in the sense that it is based on water-based predictive optimization with environmental parameters in one system.

It also assists organizations in complying with the new regulations that are coming up regarding transparency, especially regarding resources.

6. ANALYSIS AND DISCUSSION

The incorporation of predictive analytics in AI data center operations holds significant potential for water efficiency and sustainability. This section presents the expected outcome of the proposed model, in addition to the operational, environmental, and policy implications.

6.1 Anticipated Outcomes of Predictive Water Optimization

By making an accurate prediction of the water demand based on the workload and environmental conditions, the data centers are able to prepare themselves for the peak demand period and make strategies accordingly to meet the demand. It has also been researched that the management of infrastructure through data-driven approaches has the potential to improve the performance of the data centers by minimizing the inefficiencies of the systems and maximizing the potential of the resources (Cerquitelli et al., 2020; Lei et al., 2025).

In order to demonstrate the relevance of this framework, let us consider the following example. Suppose that the load in computational terms is expected to go up by 40% during peak working hours when the ambient temperature is high. It is possible to make a prediction using the proposed model that the demand for cooling will consequently increase proportionally, resulting in an increased need for water, based on the known correlations between the computational workload, heat production, and cooling needs (Gao et al., 2012; Masanet et al., 2020). Such a forecast allows one to optimize the cooling process in order to avoid unnecessary waste of water.

6.2 Operational Benefits and Efficiency Improvements

As far as operation is concerned, there are numerous advantages related to taking a predictive approach towards water management. Firstly, there is increased efficiency of the cooling process as the amount of available resources is correlated with real demand. Secondly, there is the ability to manage workloads in a better way by correlating workloads and resource availability.

One of the most considerable advantages of a predictive approach toward water management is improved reliability of the system itself. It is possible to detect potential problems with the system in time and prevent any future failures related to abnormal consumption of water or inefficiency of the cooling process (Sarker, 2021). This kind of advantage is already being achieved in practice with regard to large-scale infrastructure systems as there were improvements achieved in terms of efficiency and reducing wastage thanks to predictive maintenance and optimization (Cerquitelli et al., 2020).

In comparison to other traditional approaches that rely on static cooling systems, the predictive one promises greater efficiency due to correlation of resource availability with demand.

In reality, major tech companies like Google are already integrating water stewardship policies within their data centers, for example, using recycled water and climate-conscious cooling systems (Google, 2021). Nevertheless, this process is usually passive or policy-based rather than being predictive in nature. In contrast, our framework builds upon these initiatives to allow active and data-driven management of water consumption.

Table 3. Key Benefits of Predictive Analytics for Sustainable Water Management

Area	Traditional Management	Predictive Analytics Approach	Expected Outcome
Cooling Operations	Reactive cooling adjustments	Forecast-based adaptive cooling	Reduced water wastage
Resource Allocation	Static provisioning	Dynamic workload optimization	Improved operational efficiency
Leak Detection	Manual monitoring	AI-driven anomaly detection	Faster issue identification
Sustainability Monitoring	Periodic reporting	Real-time KPI tracking	Better ESG compliance
Water Usage Planning	Short-term operational focus	Long-term predictive forecasting	Improved sustainability planning
Risk Management	Reactive maintenance	Early warning systems	Reduced operational failures

6.3 Environmental and Community-Level Implications

The environmental benefits of predictive water optimization are significant. Water savings directly contribute to the preservation of water resources. Water preservation is vital for areas where there is a water deficit. Data centers can reduce their footprint by reducing excessive water use. It also helps data centers meet global sustainability goals.

Data centers also contribute to community welfare by reducing conflicts between community and industrial water use. In areas where there is a water deficit, data centers contribute to excessive water use. Predictive analytics helps data centers use water more responsibly by aligning data center practices with regional water availability (Mekonnen & Hoekstra, 2016; UNESCO, 2023). Ethical considerations also arise from data center operations. Data center operations must not disproportionately impact the community, particularly where there is a water deficit. Industry reports indicate that hyperscale data center operators are increasingly adopting water stewardship strategies, including recycled water usage and climate-aware cooling systems, to reduce environmental impact (Google, 2021).

6.4 Policy and Regulatory Perspectives

The environmental impact of data centers has been increasing. This issue has caught the attention of policymakers and different regulatory bodies. In addition, data analytics can be used to assist policymakers by providing evidence of resource consumption. This will enable them to establish benchmarks and ensure transparency (Haddout & Kurniawan, 2026).

The integration of these systems is in line with corporate sustainability initiatives. This allows a corporation to achieve Environmental, Social, and Governance (ESG) goals through quantified results.

The development of improved regulations, such as the mandatory disclosure of water usage and the implementation of sustainability compliance regulations, is important for ensuring accountability and adherence to the use of sustainable resource management practices.

This is especially important, as in the current scenario, the government is making efforts to ensure that environmental sustainability indicators are incorporated in regulations pertaining to the development of digital infrastructure. Quantified sustainability indicators, such as reduced WUE, are important for enforcing regulations.

6.5 Trade-Offs Between Performance, Cost, and Sustainability

Although there are benefits, there are trade-offs that are associated with the implementation of predictive analytics in the management of water in data centers. This includes the cost of the infrastructure that is required for the development of predictive analytics. Additionally, there are trade-offs that are associated with performance and sustainability. There are chances that reducing water consumption may have a trade-off on the performance of the cooling system.

However, in the future, it is apparent that the benefits of efficiency, reduced operation expenses, and sustainability will surpass the cost of investment. The sophisticated optimization tools, such as hybrid learning, are likely to achieve the optimal balance between efficiency and water conservation (Qi et al., 2023).

7. PRACTICAL IMPLICATIONS AND RECOMMENDATIONS

The practical use of the predictive analytics for the water management strategy in the AI data centers is likely to have a significant impact on the operators and the policymakers. The following section provides the recommendations on the practical use of the sustainable water management strategy.

7.1 Guidelines for Data Center Operators

The data centers need to focus on the practical use of the predictive analytics strategy. This could be achieved through the use of IoT technology, which could help the data centers monitor the temperature and water usage in real-time. This could help the data centers make predictions regarding the future (Shah et al., 2021).

The data centers also need to focus on the practical use of the machine learning strategy. The data centers need to consider the type of machine learning strategy that could be used in the data centers. The strategy could include the use of time series forecasting for predicting the demand and the use of anomaly detection for predicting the future. The machine learning strategy also needs to be updated for the purpose of accuracy (Suresh Babu C. V, Sathiya Narayana S. 2024).

In addition to this, it is also important to optimize the cooling approaches by dynamically configuring the systems based on the predictions. This will involve choosing the right cooling approaches, such as liquid cooling or hybrid cooling approaches, depending upon the nature of the workload and the environment (Zhang et al., 2021). This may also involve initial investments in infrastructure and data integration systems. This may pose challenges to small facilities. This can result in improved operational efficiency, including reduced water consumption.

In terms of implementation issues, there is a need for initial expenses regarding setting up of infrastructure, such as IoT sensors, data pipelines, and model training (Shah et al., 2021). Furthermore, the implementation of this strategy will entail dealing with some operational issues like the reliability of data, latency in real-time systems, and model maintenance. Nevertheless, these expenses can be recouped through increased efficiencies in resources, water savings, and system reliability (Cerquitelli et al., 2020).

7.2 Strategic Planning for Water-Stressed Regions

In addition, for those operating in water-scarce environments, it is important to have a strategic plan in place. In such environments, it is important to consider the availability of water in the region for the next generation. Climate conditions and sustainability factors are also important considerations when selecting a region for a data center. This is especially important when operating in environments where recurrent droughts are a problem.

In addition, it is important to have adaptive strategies in place for the cooling of the data center to ensure that it does not rely heavily on water when it becomes scarce. This will help in reducing the impact on the environment while promoting a more sustainable approach to natural resources (Mekonnen & Hoekstra, 2016).

7.3 Policy Recommendations for Water Reporting and Transparency

However, it is essential that policymakers come up with standardized frameworks for monitoring and reporting water usage in the data center. Moreover, policymakers should make it mandatory for the reporting of water usage in the data center, for example, Water Usage Effectiveness, in order to increase the level of transparency.

The regulatory bodies should ensure that there is a mandatory reporting of water usage, sustainability thresholds, as well as the imposition of sanctions for non-compliance. Furthermore, the regulatory bodies should encourage the promotion of data-driven tools for the optimization of resources, as well as the need for research on water-efficient technology.

However, there is a need for the stakeholders in the industry as well as the relevant authorities to come up with policies that take into consideration the growth of technology as well as the sustainability of the environment. Moreover, the use of standardized metrics will be essential in the monitoring of water efficiency in the data center.

There is an increasing trend toward mandatory reporting of sustainability, including the usage of water in the digital infrastructure sector (UNESCO, 2023). Regulatory changes have resulted in greater demands for reporting in relation to resource use, especially where water scarcity is an issue (Mekonnen & Hoekstra, 2016). Predictive analytics can assist businesses to remain compliant while enhancing operational efficiency.

7.4 Embedding Predictive Sustainability into Infrastructure Planning

To ensure sustainability, there is a need to integrate predictive analytics into the design and planning phases of data center infrastructure. This would require integrating the predictive models into digital twins and simulators to assess the use of resources during the design phase.

Sustainability metrics should also be integrated into the planning of data center infrastructure. In this way, data center systems would not only be effective but also sustainable.

The application of a predictive and data-based approach for the management of water resources will aid in the transition of data centers toward a more sustainable approach (Magotra et al., 2023). This is because it will ensure that the development of AI infrastructure is aligned with the global approach to sustainability.

The incorporation of sustainability metrics by top tech companies in their infrastructure planning is a clear indicator of a shift toward the adoption of a predictive approach for the management of resources by the entire industry.

8. FUTURE RESEARCH DIRECTIONS

While there is promise in the approach that is being taken with predictive analytics in terms of optimizing water management in AI data centers, there are a number of areas that need to be researched in order to make the approach more effective.

8.1 Digital Twin Technologies and Simulation-Based Optimization

An area of interest that may lead to further research would be that of integrating digital twin technologies in the framework of the predictive analytics. In essence, digital twins refer to a type of technology which can help data centers create an entire virtual representation of their infrastructure. The creation of such a virtual representation helps data centers run simulations and study the effect different scenarios have on the management of water resources. Moreover, simulations may also be carried out in order to study the behavior of the systems in case of exposure to extreme environmental conditions.

8.2 Comparative Studies Across Regions

Comparative studies in different regions should be conducted in order to understand how environmental conditions influence the usage of water. Such studies will also be useful in providing information about the effectiveness of predictive models under different climatic conditions. They can also be instrumental in helping formulate new optimization techniques by including data in different climatic conditions in the existing models. Moreover, comparative studies can help reveal the differences in strategy adopted by different regions and countries when it comes to data center operations.

8.3 Integration of Models for Optimal Use of Renewable Energy and Water Efficiency

Apart from the points highlighted in the above, there could also be consideration of integrating models for achieving optimal use of water along with renewable energy. The reason behind this is the fact that there exists an interdependence between the consumption of energy and water in data centers. As far as the future of research is concerned in this particular domain, predictive analysis of energy and water will need to be undertaken (Cina et al., 2025).

8.4 Development of Global Water Efficiency Standards

The absence of standardized metrics and benchmarking for the use of water in data centers has been identified as a challenge to the sustainability of water. Therefore, future research is required to be conducted to create universally accepted metrics and standards for the use of water in data centers, which are beyond the Water Usage Effectiveness (WUE) commonly used. This has to be environmentally friendly and consider the availability of water in the region and its sustainability (Mytton, 2021).

In conclusion, for the predictive analytics of the water management system in data centers to advance, there is a need to conduct future research to bridge the gap between artificial intelligence and environmental science to bridge the gap in water management and infrastructure engineering, which is crucial in developing a sustainable artificial intelligence data center ecosystem.

8.5 Limitations of the Study

There are various limitations associated with this analysis. To begin with, the framework presented is conceptual in nature and does not incorporate any practical data sets. Secondly, the simulation is an example of what might occur rather than a precise quantifiable model of actual events. The third limitation is that the model simplifies the interconnections among various variables. Other limitations include the variability in infrastructure, different policies in different regions, and financial limitations. Future studies will aim at validating the proposed framework with practical data sets.

9. CONCLUSION

9.1 Summary of Key Insights

The rapid development of data centers through the use of AI has posed a number of challenges, particularly with regards to the use of resources such as water. This study has identified that water consumption by data centers varies depending on certain conditions such as workload, environment, and design. It is imperative to note that reactive strategies are not sufficient to address the issue of water consumption by data centers; therefore, data centers must consider more advanced strategies.

Water consumption by data centers, particularly through AI, is a major threat to water stress and environmental degradation if not addressed immediately.

The use of predictive analytics has been identified as a potential solution to this challenge, where the data centers will be able to make accurate predictions on the consumption of water, thus addressing the inefficiencies in the data centers using machine learning and integrated systems. It means that the data centers have the ability to move from static resource management to intelligent resource management using predictive analytics, thus the need for the data centers to consider this strategy, especially because the challenges will only be on the increase over time due to the increased AI workloads, thus the resources consumed.

9.2 Contribution to Sustainable AI Infrastructure Development

This paper contributes to the body of knowledge by proposing a framework for integrated predictive analytics that focuses on addressing water sustainability concerns for AI data centers, a dimension that has not received adequate attention in previous studies.

This paper also bridges the knowledge gap between AI and environmental sustainability by showing how AI-based concepts can be used to manage vital resources. It also underscores the need to consider environmental sustainability while designing digital systems, thus promoting the development of a sustainable AI ecosystem. It integrates concepts of predictive analytics with sustainability concepts to form a foundation for future studies and implementation of resource-efficient data centers.

9.3 Toward Responsible Digital Growth

However, as digital technologies continue to improve and develop, the importance of the sustainability of the supporting infrastructure also becomes more crucial. The findings of the present study have also emphasized the importance of achieving a balanced state of development in which the advancement of technologies is linked to the management of resources.

In addition, the collaboration of various stakeholders in the industry, as well as the involvement of experts and government agencies, is also essential in the development of norms and the promotion of transparency in the implementation of sustainable infrastructure.

To conclude, the present study has demonstrated the potential of predictive analytics in the development of an efficient and responsible water management system for AI data centers. In fact, the implementation of predictive analytics can pave the way for the achievement of the long-term objective of sustainable development in the digital world. This is essential in the development of the benefits of AI without compromising the integrity of the environment.

REFERENCES

Cerquitelli, T., Nikolakis, N., Bethaz, P., Panicucci, S., Ventura, E., Macii, E., Andolina, A., Marguglio, K., Alexopoulos, K., Petrali, P., Pagani, A., van Wilgen, M., & Ippolito, M. (2020). Enabling predictive analytics for smart manufacturing through an IIoT platform. *IFAC-PapersOnLine*, *53*(3), 179–184. Advance online publication. DOI: 10.1016/j.ifacol.2020.11.029

Dayarathna, M., Wen, Y., & Fan, R. (2016). Data center energy consumption modeling: A survey. *IEEE Communications Surveys and Tutorials*, *18*(1), 732–794. DOI: 10.1109/COMST.2015.2481183

El Hanjri, M., Kabbaj, H., Kobbane, A., & Abouaomar, A. (2023). Federated learning for water consumption forecasting in smart cities. arXiv. /arXiv.2301.13036 DOI: 10.1109/ICC45041.2023.10279576

Gao, P. X., Curtis, A. R., Wong, B., & Keshav, S. (2012). It's not easy being green. In *Proceedings of the ACM SIGCOMM Conference*. DOI: 10.1145/2342356.2342398

Google. (2021). Water stewardship in data centers. https://sustainability.google/reports/google-water-stewardship

Haddout, S., & Kurniawan, T. A. (Eds.). (2026). *Handbook of Artificial Intelligence in Water Management: Towards Net Zero Emissions* (1st ed.). CRC Press., DOI: 10.1201/9781003547099

Islam, S., Smith, K., Klimes, M., & Salzberg, A. (Eds.). (2025). *Routledge Handbook of Water Diplomacy* (1st ed.). Routledge., DOI: 10.4324/9781003178439

Jiang, F., Duan, C., & Chen, B. (2025). Facility-level energy-driven water footprint and scarcity implications of Chinese data centers: A bottom-up analysis and scenario-based projection. *Applied Energy*, *399*, 126522. DOI: 10.1016/j.apenergy.2025.126522

Kollar, A., & Grady, C. (2025). The relationship between data centers and the climate is a systems challenge: A spatial analysis of United States data centers. *Environmental Research Communications*, *7*(11), 111005. DOI: 10.1088/2515-7620/ae193a

Koomey, J. (2011). *Growth in data center electricity use 2005 to 2010*. Analytics Press.

Kurniawan, T. A., & Anouzla, A. (Eds.). (2026). *Handbook of Digitalization and Big Data in the Water Sector: Emerging Opportunities and Challenges in Net Zero Emissions* (1st ed., Vol. 2). CRC Press., DOI: 10.1201/9781003536079

Lei, N., Lu, J., Shehabi, A., & Masanet, E. (2025). The water use of data center workloads: A review and assessment of key determinants. *Resources, Conservation and Recycling, 219*, 108310. DOI: 10.1016/j.resconrec.2025.108310

Leon, L., Chaplot, B., & Solomon, A. (2020). Water consumption forecasting using soft computing: A case study. *Water Science and Technology: Water Supply, 20*(8), 3576–3584. DOI: 10.2166/ws.2020.273

Liu, Y., Wei, X., Xiao, J., Liu, Z., Xu, Y., & Tian, Y. (2020). Energy consumption and emission mitigation prediction based on data center traffic and PUE for global data centers. *Global Energy Interconnection, 3*(3), 272–282. DOI: 10.1016/j.gloei.2020.07.008

Magotra, B., Malhotra, D., & Dogra, A. K. (2023). Correction to: Adaptive Computational Solutions to Energy Efficiency in Cloud Computing Environment Using VM Consolidation. *Archives of Computational Methods in Engineering, 30*(5), 3485. DOI: 10.1007/s11831-022-09868-8

Masanet, E., Shehabi, A., Lei, N., Smith, S., & Koomey, J. (2020). Recalibrating global data center energy-use estimates. *Science, 367*(6481), 984–986. DOI: 10.1126/science.aba3758 PMID: 32108103

Mekonnen, M. M., & Hoekstra, A. Y. (2016). Four billion people facing severe water scarcity. *Science Advances, 2*(2), e1500323. DOI: 10.1126/sciadv.1500323 PMID: 26933676

Mytton, D. (2021). Data centre water consumption. npj. *NPJ Clean Water, 4*(1), 1–8. DOI: 10.1038/s41545-021-00101-w

Qi, S., Milojicic, D., Bash, C., & Pasricha, S. (2023). SHIELD: Sustainable hybrid evolutionary learning framework for data center optimization. arXiv. https://doi.org//arXiv.2308.13086DOI: 10.48550

Rongon, R. K., & Das, K. (2025). Energy-aware data center management: A sustainable approach to reducing carbon footprint. arXiv. https://doi.org//arXiv.2509.10462DOI: 10.48550

Sarker, I. H. (2021). Machine learning: Algorithms, real-world applications and research directions. *SN Computer Science, 2*(3), 160. DOI: 10.1007/s42979-021-00592-x PMID: 33778771

Shah, A., Fauzi, S., Gining, R. A., Razak, T. R., Jamaluddin, M. N. F., & Maskat, R. (2021). IoT-based monitoring systems for smart cities. *Indonesian Journal of Electrical Engineering and Computer Science, 21*(1), 450–456. DOI: 10.11591/ijeecs.v21.i1.pp450-456

Suresh Babu, C. V. Subha Sree, Prahadeesh, V., Adams, S. R., & Prem Kumar, S. (2024). IoT-Driven Water Management Solutions for Sustainable Agriculture in the Age of Autonomy. In R. Rajappan, N. Gowri Ganesh, J. Daniel, A. Ahmad, & R. Santhosh (Eds.), The Convergence of Self-Sustaining Systems With AI and IoT (pp. 242-260). IGI Global. DOI: 10.4018/979-8-3693-1702-0.ch013

Suresh Babu, C. V. Sathiya Narayana S. (2024). Navigating Uncharted Waters: Overcoming Challenges in Integrating Machine Learning and Deep Learning for Enhanced Clinical Risk. In H. Liu, R. Tripathy, & P. Bhattacharya (Eds.), Clinical Practice and Unmet Challenges in AI-Enhanced Healthcare Systems (pp. 289-310). IGI Global. DOI: 10.4018/979-8-3693-2703-6.ch015

Suresh Babu, C. V., Yadav, S., Ganesh, S., Niraianban, A. R., & Barathi, A. (2025). Logistic Regression-Based Injury Prediction and Prevention in Sports Analytics: An AI Approach. In Choudhury, T., Arya, P., Kotecha, K., Sharma, A., & Um, J. (Eds.), *AI and Machine Learning Applications in Sports Analytics* (pp. 103–114). IGI Global Scientific Publishing., DOI: 10.4018/979-8-3693-5385-1.ch005

UNESCO. (2023). United Nations world water development report 2023. https://www.unwater.org/publications/un-world-water-development-report-2023

Zhang, Q., Meng, Z., Hong, X., Zhan, Y., Liu, J., Dong, J., Bai, T., Niu, J., & Deen, M. J. (2021). A survey on data center cooling systems: Technology, power consumption modeling and control strategy optimization. *Journal of Systems Architecture*, *119*, 102253. Advance online publication. DOI: 10.1016/j.sysarc.2021.102253

Chapter 2

The Impact of Artificial Intelligence Data Centers on Global Water Resources and Management:

A Review and Conceptual Framework for Smart Water Governance

Usman Abidemi Sarumi
http://orcid.org/0009-0003-0784-8055

AI and Software Engineering Dept., AI and IoT Research Center, AI and Informatics Faculty, Near East University, Mersin, Turkey

Osman Abdalla Khalfalla Alhussein
http://orcid.org/0000-0002-6248-2530

AI and Software Engineering Dept., AI and IoT Research Center, AI and Informatics Faculty, Near East University, Mersin, Turkey

Sinem Alturjman

AI and Software Engineering Dept., AI and IoT Research Center, AI and Informatics Faculty, Near East University, Mersin, Turkey

Hitesh Mohapatra
http://orcid.org/0000-0001-8100-4860

School of Computer Engineering, Kalinga Institute of Industrial Technology (Deemed to be University), Bhubaneswar, India

DOI: 10.4018/979-8-2600-2761-5.ch002

Fadi Al-Turjman
http://orcid.org/0000-0001-6375-4123
AI and Software Engineering Department, AI and IoT Research Center, AI and Informatics Faculty, Near East University, Mersin, Turkey

ABSTRACT

Due to global demand for advanced technology that requires high-performance computational infrastructure, there has been rapid growth of artificial intelligence (AI) usage. Among this infrastructure are AI datacenters, which consume a large amount of electricity and water. As AI models grow more resource-intensive, the need for water power generation, chip manufacturing, and cooling has risen as critical costs that are overlooked. The paper adopts a critical review of existing literature with a conceptual framework for smart water management in the context of AI-intensive urban infrastructure. The study evaluates how rapid AI development intersects with proposed strategies and freshwater scarcity for reducing water footprints through policy action and technological innovation. The findings highlight an urgent need for transparent water-use reporting, region-specific regulation and sustainable cooling solutions to mitigate the escalating pressure of AI infrastructure on global water resources

1. INTRODUCTION

Water is a crucial resource that sustains life, supports economic development, and helps protect the environment. Water is essential to all living things for life-sustaining processes, and it supports the many different economies and communities that exist around the planet. However, quite a few metropolitan regions are confronting severe water shortages, losing capital through decaying older systems while also contaminating the existing water supply. The supply of quality and safe water is on the rise, but the distribution processes are still dependent on old technology and over-reliant on the scant supply of fresh water. This set of challenges is aggravated by conventional approaches to water resource management and control through a strict centralized system of water governance and infrastructure that is often inefficient at coping with the fast-changing requirements of cities. Scarcity of water resources makes it more obvious that new ways to manage them thoughtfully and make them available for future generations are increasingly needed as shown in Figure 1.

"Recent smart-water research emphasizes that climate change, urban growth, aging infrastructure, and increasing demand require cities to move beyond traditional centralized water management toward integrated sensor-based monitoring,

digital control, and adaptive governance (Oberascher et al., 2022; UNESCO World Water Assessment Programme, 2024). "

Figure 1. Global distribution of physical and economic water scarcity

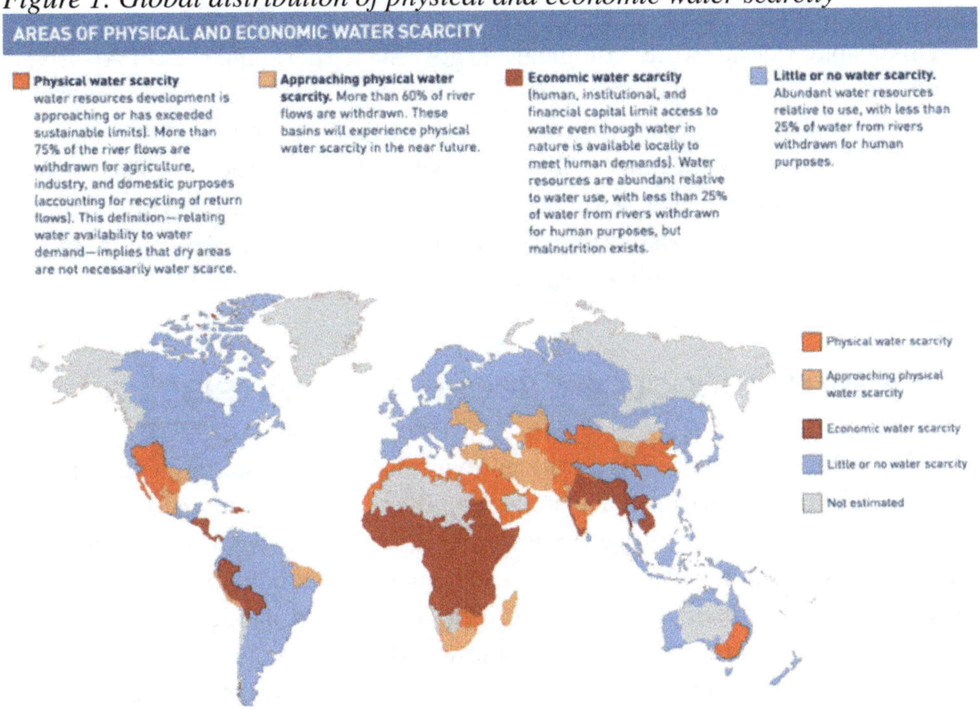

AI has become a major element in the growth and development of useful applications across different domains, with implementations spanning from self-learning and natural language processing to medical assessment, artificial intelligence has emerged as a key element of modern technology. High-performance IT infrastructure is becoming increasingly necessary as AI capabilities grow. Contemporary AI data centers house thousands of servers, machine-learning accelerators, and high-density GPU clusters in contemporary AI data centers. These devices produce substantial heat and require reliable cooling systems to remain stable (Mytton, 2021).

Water consumption has recently come to light as a crucial environmental factor, even though energy use has long dominated conversations about technology viability. Water is used directly for evaporative cooling and indirectly for the production of semiconductors and power. This need is increased by the growing size and complexity of AI models, such as transformer frameworks and massive base algorithms (International Energy Agency, 2025). If the current strain on freshwater resources continues along with no remedial action being taken, the depletion of natural water resources would be made worse in the long run. The implementation of smart water management solutions is critical to endure these system shocks while ensuring water system reliability. The future is deterministic. Thus, societies willing to embrace change and technology today stand a chance to protect their water resources and provide for a better tomorrow. This situation is as shown in figure 2.

This chapter examines the ecological and governance consequences of AI-driven data centers, with particular emphasis on the pressure they place on global water resources. It also identifies mitigation strategies that can reduce environmental impacts and support environmentally responsible AI infrastructure. The chapter proceeds as a critical review and conceptual analysis, connecting AI data center growth with smart water management, circular economy principles, digital twins, regional water stress, and policy design.

2. REVIEW METHODOLOGY, SCOPE, AND CONTRIBUTION

This chapter is positioned as a critical review and conceptual framework rather than an experimental performance study. The literature was examined through an integrative lens that combines peer-reviewed research on data center energy and water demand, smart water management, urban resilience, semiconductor manufacturing, and corporate environmental reporting. Priority was given to studies and reports that explicitly address water consumption, water withdrawal, water usage effectiveness (WUE), energy-water interactions, cooling technology, AI infrastructure, and governance mechanisms. The review also considers practitioner reports because data center operators often disclose water-related information through sustainability documents before it appears in academic databases.

The scope of the chapter is intentionally lifecycle-based. A facility-level analysis would focus only on evaporative cooling or mechanical cooling inside the data center boundary. Such a narrow boundary can understate the real water implications of AI infrastructure. For this reason, the chapter evaluates three water-demand channels: direct operational water used on-site, indirect water embedded in electricity generation, and upstream water embedded in semiconductor production, server manufacturing, construction materials, and supply chains. This boundary is important because an operator may reduce on-site water consumption while increasing electricity demand or shifting impacts to water-stressed manufacturing regions.

The original contribution of the chapter is the development of a practical conceptual framework for AI data center water governance. The framework links AI compute growth, infrastructure design, lifecycle water demand, regional water stress, smart mitigation technologies, governance instruments, and measurable sustainability outcomes. It is designed for researchers, smart-city planners, water utilities, data center operators, and policymakers who need a structured way to evaluate the water consequences of AI infrastructure. The framework does not treat AI only as a source of environmental burden; it also examines how AI, digital twins, and anomaly detection can improve water conservation when they are deployed responsibly.

The review is guided by four research questions. First, through which direct, indirect, and embedded pathways do AI data centers affect freshwater resources? Second, which smart technologies and circular-water strategies can reduce water demand without creating hidden energy or equity trade-offs? Third, how do regional climate and governance differences shape the severity of AI-related water impacts? Fourth, which policy mechanisms can improve transparency, accountability, and responsible siting of AI infrastructure? These questions provide the analytical structure for the remainder of the chapter.

3. THE GROWTH OF AI DATA CENTERS AND THE AI-WATER NEXUS

The global deployment of data centers is increasing due to the growth of AI. To accommodate deep learning operations, continuous decision-making, and extensive training, cloud service providers and tech firms are developing specific AI facilities. Recent projections indicate that the demand for extremely dense computational workstations will be a major factor in the potential doubling of the worldwide AI computing industry by 2030 (International Energy Agency, 2026).

A significant portion of the electricity used by data centers can be generated by water-intensive thermal power plants. Water is needed for cooling, steam production, and thermal waste distribution in coal, natural gas, and nuclear power plants (Macknick et al., 2012). This indirect water impact is increased by AI operations, which can necessitate weeks of nonstop retraining.

3.1 Conceptual Framework for AI Data Center Water Sustainability

The conceptual framework proposed in this chapter begins with the observation that AI compute demand is not environmentally neutral. As models become larger and more widely deployed, compute demand expands across training, fine-tuning, inference, storage, and network services. This demand is translated into physical infrastructure through high-density server halls, accelerators, power distribution equipment, cooling systems, backup power systems, and water-related support facilities. The water footprint then emerges through three linked routes: direct operational water consumption, indirect water use associated with the electricity mix, and embedded water use in hardware production and construction. Figure 2 presents this relationship as a governance-oriented framework rather than a purely technical cooling diagram.

The framework emphasizes that sustainability outcomes depend on the interaction between technology and place. A cooling design that performs well in a cool and water-abundant region may be inappropriate in a hot, arid, or drought-prone watershed. Likewise, a water-efficient cooling system can increase energy demand if it relies heavily on mechanical refrigeration. Responsible decision-making therefore requires simultaneous evaluation of WUE, power usage effectiveness (PUE), carbon intensity, local water stress, community water needs, and long-term climate projections. The framework also recognizes that smart mitigation tools, including leak detection, predictive maintenance, digital twins, and reclaimed water systems, must be combined with transparent governance to be socially credible.

Figure 2. Conceptual framework linking AI compute demand, data center operations, lifecycle water demand, regional water stress, smart mitigation, governance, and sustainability outcomes.

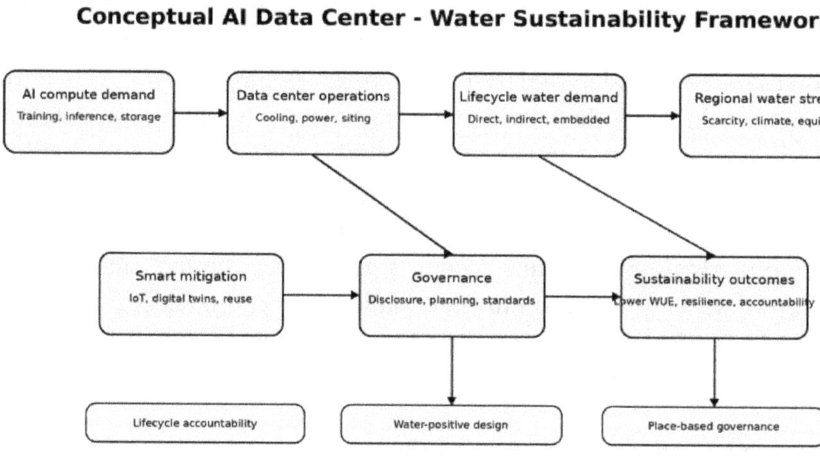

3.2 Lifecycle Water Demand Pathways in AI Infrastructure

A lifecycle interpretation is necessary because AI infrastructure can shift water pressure across locations and sectors. The following matrix summarizes the main pathways that should be evaluated when assessing the water implications of AI data centers as presented on Table 1.

Table 1. Lifecycle water-demand pathways relevant to AI data center sustainability assessment.

Pathway	Water mechanism	Main risk	Sustainability response
Direct operational water	Evaporative cooling, humidification, water treatment, blowdown, domestic uses	Local consumption can compete with municipal, agricultural, or ecological water needs.	Use reclaimed or non-potable water, closed-loop cooling, optimized WUE, dry/ hybrid cooling where appropriate, and transparent facility-level reporting.
Indirect electricity-related water	Thermal power plants consume and withdraw water for steam generation and cooling.	A low on-site WUE can still hide high water demand if electricity comes from water-intensive generation.	Evaluate source-grid water intensity, procure low-water renewable electricity, and coordinate siting with grid and water plans.
Embedded hardware water	Semiconductor fabrication and hardware manufacturing require ultrapure water and complex chemical processes.	Impacts may occur outside the data center host community and remain invisible in local permitting.	Include supply-chain water disclosure, supplier standards, circular hardware policies, and longer equipment lifecycles.
Construction and land-use water	Concrete, cooling towers, landscaping, dust control, and site preparation require water.	Rapid cluster development can strain watersheds before operations begin.	Require water impact assessments, drought-resilient landscaping, stormwater capture, and cumulative regional planning.

4. SMART TECHNOLOGY AND WATER MANAGEMENT IN DATA CENTER-ADJACENT SMART CITIES

Smart solutions for water management enable smart cities to tackle water issues while ensuring the resources are used in a fair and sustainable fashion. Figure 3 presents a case of automated control systems for effective water management, smart solutions that allow real-time monitoring and predictive analytics are key.

Figure 3. Representative illustration of smart water management

4.1 Implementation of IoT and Sensor Technologies

Strategic placement of IoT-enabled sensors at water supply reservoirs, sewage treatment plants and water distribution pump stations ensure that water quality and supply pressure and flow rate are met. This enables monitoring of the water distribution system for leakages, contamination as well as operational faults (Oberascher et al., 2022). The data is collected and analyzed continually. Proactive automated leak detection systems enable intervention that improves accuracy and minimizes water loss and infrastructure damage. Smart IoT-based water distribution networks have improved the efficiency of real-time monitoring and leak detection over the last few years. In some metropolitan areas, non-revenue water losses were reduced by up to thirty percent. (Oberascher et al., 2022).

4.2 AI Tools for Predictive and Reactive Water Intelligence

AI systems can use historical and real-time water-use data to predict demand shifts, identify potential supply deficits, and support planning for more stable supply curves. Machine learning algorithms make it possible for the municipalities to optimize water consumption at a macro level because they are able to follow consumption patterns (Salem et al., 2024). All-in-one AI platforms can also provide assistance to the management of waste water treatment plants by enabling less chemicals to be used, less energy to be consumed, and more purification to be done.

4.3 Water Trading, Billing, and Accountability Through Blockchain Technologies

Trading, billing, and resource management can be improved through blockchain technology because it can increase transparency and accountability (Zhou et al., 2020). The cities can create optimal mechanisms for the pricing of water through the management of decentralized smart contracts, blockchain management practices, and blocking of illegal water use. Incorporating blockchain water management technology can facilitate accountability on water consumption billing, restrict unauthorized usage, and enhance water trade between city administrations. (Zhou et al., 2020).

4.4 Efficient Water Distribution Through Smart Infrastructure

Computer-controlled networks for distribution of water ensure that all areas of a city receive water according to demand by varying the rate of flow and the pressure level of water supplied to them.

Smart meters allow monitoring of water usage at the level of households, business entities, and industrial sites to be done live by both the users and the officials. The use of AI analytics in combination with smart valves and flow rate controllers improves the situation with the distribution of water in times of high demand (Zhou et al., 2020).

4.5 Semiconductor Manufacturing for AI Data Center Devices

Semiconductor fabrication plants produce AI accelerators such as GPUs, TPUs, and NPUs. To rinse silicon wafers, eliminate contaminants, and uphold regulated clean-room conditions, these establishments use millions of liters of ultrapure water every day (Wang et al., 2023). Data center impact studies sometimes do not include the water footprint associated with chip production.

5. RECYCLING, REUSE, AND THE CIRCULAR WATER ECONOMY IN SMART CITIES

The implementation of technologies focused on the recycling and reuse of water is key in managing sustainable water resources in smart cities. Considering the limits on freshwater resources, cities should evolve from conventional water supply systems to circular water economies.

Sustainable urban water management increasingly depends on circular water systems, decentralized reuse, smart meters, and sensor-based monitoring that reduce pressure on freshwater resources while improving operational resilience (Oberascher et al., 2022; UNESCO World Water Assessment Programme, 2024). "

5.1 Wastewater Treatment and Reuse

Advanced wastewater treatment facilities are employing membrane bioreactors (MBRs) alongside reverse osmosis and ultraviolet (UV) disinfection to purify sewage water to drinking water standards. Industrial and power plants as well as agricultural industries can use treated wastewater for cooling, irrigation, landscaping or even potable water (UNESCO World Water Assessment Programme, 2024).

Cities like Los Angeles and Singapore have large-scale water recycling initiatives like Pure Water San Diego and NEWater, respectively, which show the safe reuse of treated water is possible.

5.2 Rainwater Harvesting and Urban Water Storage Solutions

These processes are helpful in restoring groundwater reserves, reducing reliance on drinking water, and helping manage urban flooding. Constructed wetlands, bioswales, and urban forests filter polluted water before it is discharged into rivers and aquifers (UNESCO World Water Assessment Programme, 2024).

Pressure on drainage systems is reduced and surface runoff is minimized through permeable pavements and rain gardens. This situation is as shown in Figure 4, indicating the relationship between the traditional water usage and smart system usage.

Figure 4. Graphical illustration of water savings resulting from the use of smart systems

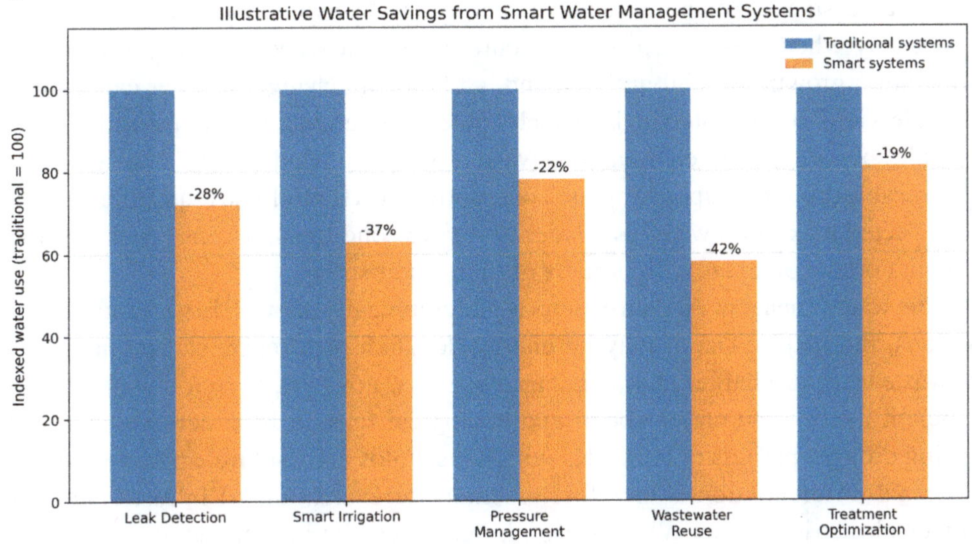

5.3 Geographical and Contextual Differentiation

Illustrative international cases show the importance of context. Singapore demonstrates how a water-stressed city-state can combine advanced water reuse, strict planning, and public communication through programs such as NEWater. Amsterdam and Copenhagen illustrate how smart monitoring, flood prediction, and canal or drainage management can connect digital systems with urban water resilience. In contrast, rapidly expanding data center clusters in arid or semi-arid regions need stronger public reporting and more explicit coordination with water utilities. These examples indicate that smart water management is not simply a technology problem; it is a planning, governance, and trust problem.

Water-abundant regions should not be treated as risk-free. Heavy rainfall, flooding, aging drainage systems, and polluted runoff can threaten the reliability of water and energy infrastructure. Data center growth in these regions should therefore be integrated with floodplain management, stormwater capture, and digital twin models of urban infrastructure. In cold climates, free cooling can reduce water demand, but the benefits depend on electricity mix, heat reuse opportunities, and the quality of environmental reporting. The best approach is place-based: environmental performance should be assessed according to the hydrological and social conditions of the host watershed.

Water-scarce regions require a precautionary approach to siting. New AI data centers in arid or drought-prone watersheds should be evaluated through cumulative impact assessments rather than isolated facility-level permits. The relevant question is not only whether one facility can secure water, but whether multiple facilities, population growth, agricultural demand, and climate change together can exceed the safe yield of the watershed. In such contexts, operators should prioritize non-potable sources, reclaimed municipal wastewater, air-cooled or hybrid designs, and enforceable drought-response plans. Local authorities should also require disclosure of expected maximum water use under heatwave conditions, because peak cooling demand often coincides with periods of high water stress.

The water impact of AI data centers is geographically uneven. In water-abundant regions, the main concern may be energy demand, heat reuse, carbon intensity, or wastewater discharge. In water-scarce regions, however, even a comparatively efficient facility can create social and ecological tension if it increases demand during drought periods or competes with households, agriculture, and ecosystems. Regional differentiation is therefore essential for evaluating AI infrastructure. A universal metric such as WUE is useful, but it must be interpreted alongside watershed stress, seasonal variability, groundwater depletion, climate projections, and local governance capacity.

Table 2. Regional water-stress and AI data center governance risk matrix.

Regional context	Dominant water concern	AI data center risk pathway	Recommended governance response
Water-scarce and drought-prone watersheds	Competition for limited freshwater, drought vulnerability, and social pressure on municipal supply	Evaporative cooling, potable-water withdrawal, and electricity-related water demand can intensify scarcity during peak heat periods	Require watershed-level impact assessment, drought contingency planning, reclaimed-water prioritization, and transparent facility-level WUE reporting
Water-abundant but flood-prone regions	Stormwater overload, flood disruption, and polluted runoff rather than absolute scarcity	Large impermeable campuses, heat discharge, grid stress, and backup infrastructure can increase local resilience risks	Integrate siting approval with flood-risk mapping, green infrastructure, emergency water planning, and climate-resilient drainage design
Rapidly urbanizing regions	Weak infrastructure, uneven access, limited monitoring capacity, and governance fragmentation	New digital infrastructure can compete with households, agriculture, and industry if planning systems are underdeveloped	Adopt phased permitting, public consultation, infrastructure co-investment, local workforce training, and minimum disclosure obligations
Regions with reclaimed-water capacity	Need to protect potable water while encouraging circular water use	Cooling demand can be shifted from potable sources to treated wastewater when quality and distribution networks permit	Create incentives for non-potable cooling, co-location with wastewater reuse systems, and standards for water quality, discharge, and reuse safety

6. PUBLIC ENGAGEMENT, GOVERNANCE, AND POLICY FRAMEWORKS

In smart cities, the efficient management of water resources is dependent on governance and public engagement, which is facilitated through modern technology. The local authority has to devise controlling measures in the form of policies, grants, and campaigns to educate residents and businesses about water conservation (UNESCO World Water Assessment Programme, 2024).

6.1 Regulatory Policies and Water Governance Models

Integrated policies to conserve water would include limitations on groundwater pumping and contamination; as well as the adoption of water-saving structures. Water governance may be relaxed to allow local actors at the community level to engage in the policymaking processes.Carefully designed campaigns for the efficient use of water resources can reasonably control overconsumption and water waste. Gamification and rewarding smart water practices can motivate people and businesses to automate water usage (UNESCO World Water Assessment Programme, 2024).

6.2 Voluntary Practices, Regulatory Mechanisms, and Reporting Standards

A stronger governance discussion must distinguish voluntary practices from regulatory mechanisms. Voluntary practices include corporate water-positive commitments, sustainability reporting, procurement of renewable electricity, restoration projects, internal WUE targets, and community engagement programs. These practices can accelerate innovation, but they depend on the operator's willingness to disclose data and maintain long-term commitments. They may also vary across regions, which makes comparison difficult for communities and policymakers.

Regulatory mechanisms include permitting requirements, mandatory water impact assessments, facility-level water reporting, restrictions on freshwater use during droughts, wastewater discharge standards, and land-use planning rules that integrate energy and water infrastructure. Regulatory instruments are important because data center clusters can create cumulative effects that are not visible when facilities are approved one by one. Urban planning authorities should therefore coordinate with water utilities, electricity regulators, and environmental agencies before approving large AI infrastructure projects.

International governance gaps remain significant. There is no universally adopted reporting standard that requires all data center operators to disclose comparable facility-level water consumption, water withdrawal, reclaimed water use, source-water type, peak demand, and watershed stress. WUE is valuable but incomplete because it does not by itself show whether water is potable or non-potable, whether use occurs during drought periods, or whether indirect electricity-related water has been included. A more complete disclosure approach would report direct water consumption, withdrawal, source type, discharge quality, energy-water intensity, and replenishment claims separately.

Policy design should also avoid shifting burdens. A ban on evaporative cooling may reduce local water use but increase electricity demand and associated carbon emissions if implemented without grid planning. Conversely, a facility with high evaporative cooling efficiency may reduce electricity use but increase local water consumption. A responsible policy framework should evaluate trade-offs using multiple indicators rather than a single metric. The most effective governance model is therefore a multi-criteria approach that combines WUE, PUE, carbon intensity, water stress, social equity, and resilience under extreme weather.

Table 3. Policy and governance mechanisms for AI data center water sustainability.

Governance mechanism	Purpose	Application to AI data centers	Expected benefit
Facility-level WUE and withdrawal reporting	Make operational water use comparable across facilities and regions	Require annual disclosure of water withdrawal, water consumption, source-water type, reclaimed-water share, and seasonal variation	Improves transparency and allows regulators to compare performance across operators
Mandatory water impact assessment	Evaluate cumulative watershed pressure before approval	Assess direct cooling water, electricity-related water, semiconductor supply-chain water, and construction impacts	Reduces hidden trade-offs and improves siting decisions before investment lock-in
Drought and heat-emergency rules	Protect critical community water needs during climate stress	Define cooling restrictions, emergency backup procedures, and alternative water sources during droughts or heat waves	Prevents AI infrastructure from worsening public water insecurity during extreme events
Urban planning authority coordination	Align data center growth with land use, grid planning, and water infrastructure	Link permitting to watershed plans, grid decarbonization plans, reclaimed-water pipelines, and stormwater infrastructure	Supports integrated governance rather than isolated corporate sustainability claims
Voluntary water-positive commitments with verification	Encourage restoration beyond legal minimums while reducing greenwashing risk	Require third-party verification of replenishment projects, watershed relevance, and additionality	Improves credibility of corporate stewardship and connects projects to local water conditions

6.3 Sustaining Water Reserves and Climate Resilience Financing

Public-private partnerships (PPPs) can be used to fund advanced technology and infrastructure for water supply and wastewater treatment for better integration and resiliency of water systems. This could also assist in lessening the burden on additional funding for water technology improvement (UNESCO World Water Assessment Programme, 2024).

Climate change and its effects such as extreme weather events and changes in precipitation patterns necessitate that cities adopt a broader approach towards water management including mitigation of climate change.

"Sustainable water management frameworks for smart cities should combine predictive analytics, decentralized decision-making, transparent reporting, and flexible infrastructure systems to strengthen resilience under water scarcity and climate variability (UNESCO World Water Assessment Programme, 2024; World Resources Institute, 2023).

There are smart solutions and cutting-edge technology to address the issues of water shortage, waste, and inefficiency within the context of smart cities, which can be defined as Sustainable Water Management. Multiple solutions exist using AI, IoT, and blockchain technology, coupled with forecasting method of analysis to optimize the resilience, effectiveness and water supply system that is sustainable.

Selected Works: Water Management Innovations from Smart Cities

Several cities globally employ advanced water management systems and serve as benchmark cities for the developing ones to follow:

- Singapore: With the implementation of the NEWater program, Singapore can now treat wastewater into drinking water which reduces the quantity of water needed to be imported.
- Copenhagen: Uses AI flood prediction and smart drainage systems to deal with climate change induced water issues.
- Amsterdam: Uses IoT-enabled systems for water quality monitoring to ensure safe drinking water and manage canals sustainably.
- Barcelona: Develops automated irrigation systems for urban agricultural parks and green areas to promote taking care of water resources.

There are smart solutions and cutting-edge technology to address the issues of water shortage, waste, and inefficiency within the context of smart cities, which can be defined as Sustainable Water Management. Multiple solutions exist using AI, IoT, and blockchain technology, coupled with a framework for forecasting purpose to focus on sustainability stability, resilience and effectiveness of water supply system (Oberascher et al., 2022; UNESCO World Water Assessment Programme, 2024).

6.4 Sustainable Water Management in Smart Cities: A Holistic Perspective

The capacity for resilience is largely dependent on the sustainable use of water resources, which are essential to the development of smart cities. The efficient use, equitable distribution, and sustainable management of water resources are severely hampered by environmental concerns brought on by population growth. Urban areas' continuous physical growth puts more strain on the current water supply infrastructure and changes the operational paradigm from traditional water management to one that is more adaptable and technology-driven. AI can forecast water consumption, identify leaks, and optimize networks of distribution by analyzing incoming data from automated sensors. Blockchain technology can be used to safeguard water activities.

These cities have the potential to develop sophisticated and flexible water management systems capable of providing resource, ecological, and climate change balance with the desired level of protection from disasters and depletion by redesigning the whole water cycle from one end to another including resource extraction, purification, distribution, consumption, and disposal and reuse.

7. THE FUNCTION OF DIGITAL TWINS IN WATER INFRASTRUCTURE

The use of digital twins, or virtual models of real water systems, which allow for continuous surveillance, modeling, and automated system enhancement, is one of the most significant changes in urban water administration.

Digital twins establish a new structure to facilitate maintenance prediction, finding flaws, and efficiency in operations by utilizing sensor networks, cloud-based computing, and machine learning-based analysis.

Stressing the need for independent decision making in resource poor settings, a reinforcement learningbased intelligent planning agent system for water supply management was proposed as a case study of decentralized system (Oberascher et al., 2022).

By combining hydrological and climate models with urban planning, digital twins promote decisions based on data. By modeling water demand, authorities may more efficiently distribute water resources and start conservation efforts before water stress occurs. Additionally, these models can take into account socioeconomic factors that aid cities in developing policies for fair water allocation.

Furthermore, self-learning techniques that gradually increase operational efficiency and predicted accuracy can be created by implementing machine learning in a digital twin. As more system-related information is gathered gradually, these mechanisms grow increasingly adept at identifying adaptable water administration solutions for recently developed issues as shown in Figure 5.

Figure 5. Integration of new digital technologies.

7.1 Decentralized Systems of Water Treatment and Distribution

Decentralizing water treatment and distribution is another aspect of sustainable urban water management. Inefficiencies in centralized systems are frequently brought on by increased energy use, water loss from massive pipelines, sluggish contamination detection, and other problems. Water safety can be strengthened and reliance on large facilities reduced by implementing flexible and compact water treatment facilities.

Solutions that use decentralized treatment like biofiltration units, membrane based purification, and constructed wetlands allow for the direct reuse of wastewater in defined areas such as districts, industrial zones or even residential areas which diminishes the demand for freshwater resources.

Decentralized systems also provide resiliency in water infrastructure by minimizing the risk of failure on a wider level. When one treatment facility stops working, localized smaller systems can continue to provide water to people. Also, decentralized networks make it easier to use other sources of water such as harvested rainwater or recycled greywater which decreases the reliance on freshwater supply, this situation is illustrated in Figure 6.

An additional advantage of using a decentralized system in the management of water is reducing the consumption of energy in long-distance transportation of water. The distribution and pumping of water use a significant amount of energy because fuels are used, and this leads to emissions of greenhouse gases. Cities can reduce cost as well as harmful effects on the environment by treating water and reusing it nearer to the location it is used. In areas where water is less abundant, decentralized methods also prevent the wasteful use of alternative sources, such as atmospheric water generators and desalination plants, without straining centralized infrastructure.

Figure 6. Illustration of decentralized water treatment and distribution.

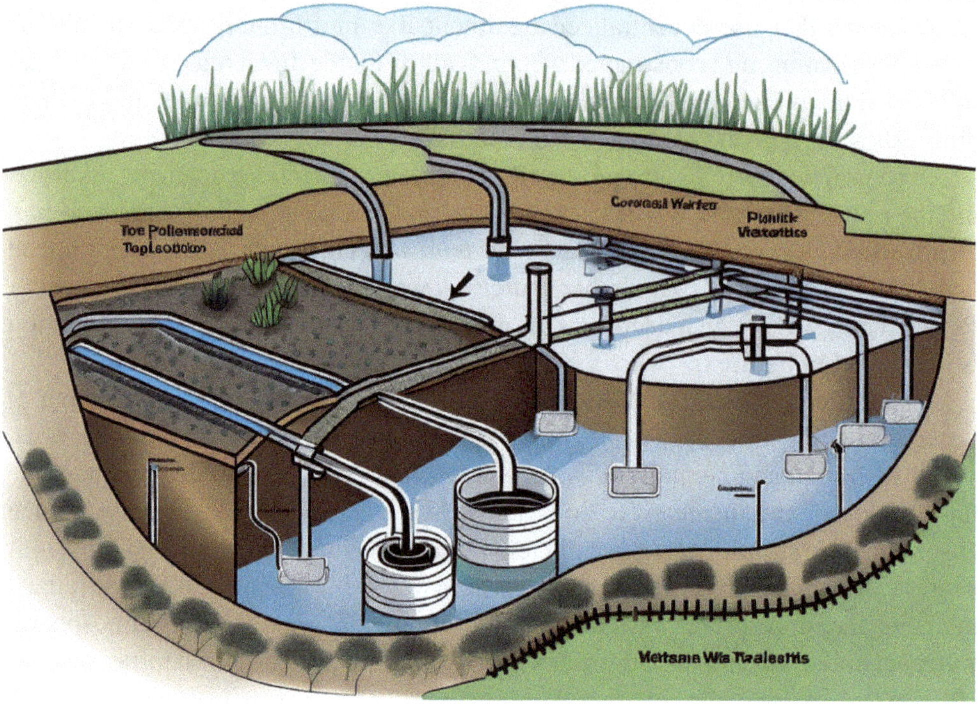

7.2 Advances in Water Purification Technologies

Progress in bioengineering and materials science has also enhanced the sustainability of water. Filtration technologies that make use of nanomaterials are able to make the process of purification much more efficient by removing minute plastic particles, heavy metals, and even residues of medicines from water. Moreover, biomimetic techniques for desalination, inspired by mangrove filtering and fish gill ion exchange for ion fluids, provide an essence for sustainable desalination without putting too much energy burden.

These techniques improve the treatment of coming contaminants such as endocrine disruptors and antibiotic resistant microbes, which have tremendous health impacts. With population growth in cities, facilitating high water quality through these systems becomes more important for public health and environmental protection.

They presented the use of Graph Neural Networks (GNNs) with water quality prediction in distribution networks with sparse monitoring and demonstrated that the spatial relationships between the nodes can greatly enhance the accuracy of the predictions. Nonetheless, their model was limited to the quality of water and did not incorporate system wide optimization as a consideration (Salem et al., 2024).

IoT sensors intelligent systems are capable of real-time monitoring of water quality and automatic adjustment of purification processes. These systems are more efficient and also safer for human consumption because there is less reliance on chemicals and lower operational expenditure. Moreover, AI systems can adjust the filtration parameters in advance and improve the efficiency of the system by anticipating changes in water quality due to industrial discharges, seasonal changes, and all the other filters.

7.3 Nature-Based Solutions for Climate Resilience

Managing water resources is a long-term strategic goal for climate resilience, in which every city will have to learn to adapt with the rising sea level, warmer temperatures and random rainfall.

Urban sponge designs, city greening, and green mosques are all examples of nature friendly designs that make use of the natural hydrological cycle for better urban planning. They will lead to improved groundwater levels, reduced surface runoff and the urban heat island effect as well as improved biodiversity and air quality. In addition, improved water resource availability makes the area less vulnerable to long dry spells and the inevitably high water scarcity events aquifer recharge and storm water harvesting programs.

Cities are now constructing multi-purpose landscapes that function as parks and aid in managing water resources. Unlike conventional wetlands, fusion pavements, and bioswales are more aesthetically pleasing. They help in holding storm water and also help in increasing biodiversity in the city. These actions aid in the overall approach for keeping water resources sustainable to the city as the city infrastructure works seamlessly with the natural water ecosystems to manage the water cycles.

Applied federated learning to water consumption prediction to enhance privacy measures. While this innovation mitigated data sharing risks, it did so at the expense of optimizing system performance in real time (El Hanjri et al., 2023).

Additionally, cities are incorporating river restoration projects and green corridors which aid in reconnecting natural water bodies with urban areas. This shift results in restoring the natural floodplain which serves as a buffer during severe climatic events. The additional preservation and restoration of mangroves and marshes as well as urban forests helps in capturing carbon and providing a habitat for wildlife. Also, helping in decreasing the temperature of the area.

For academic readers, the chapter contributes a structured review of AI infrastructure water impacts and highlights gaps in lifecycle accounting and policy analysis. For practitioners, it identifies actionable strategies such as reclaimed water use, hybrid cooling, sensor-based leak detection, regional planning, and transparent reporting. For policymakers, it clarifies where voluntary industry commitments are insufficient and where regulatory coordination is needed. This audience-oriented structure improves the chapter's flow and makes the argument more directly aligned with publication standards for a scholarly book chapter.

The proposed study theme requires the chapter to connect emerging AI infrastructure with smart-city sustainability, climate stress, and resource governance. The revised organization therefore moves from the general water scarcity problem to the specific AI data center-water nexus, then to smart mitigation technologies, circular-water strategies, governance, regional differentiation, digital twins, and prior studies. This flow is intended to serve a broad target audience that includes researchers, graduate students, city planners, water utility managers, technology managers, and policymakers. Technical concepts are explained in relation to practical water-management challenges so that the chapter remains accessible without losing analytical depth.

8. COMPARISON WITH PRIOR STUDIES

El Hanjri et al. (2023) applied federated learning to water-consumption forecasting in smart cities. Their study is useful for privacy-preserving prediction, although it does not fully address system-wide real-time optimization.

The LSTM component was able to capture water usage trends over a long period, even with sharp shifts caused by seasonal or extraordinary weather shifts. Unlike traditional time-series methods, LSTM networks managed to remember context throughout the entire sequence, offering better accuracy and more informed predictions. In the same fashion, the spatial hidden identifiers and node-level anomaly in network topologies was expertly handled by the GNN module that k-means or DBSCAN would not be able to, due to their lack of structured and flat input data handling (Hintemann & Hinterholzer, 2020).

One more benefit with our system is the ability to make decisions instantly. Decisions that can be made in an instant with the aid of reinforcement learning(RL) algorithms that learn from the system feedback. Prior systems were batch-based, where data would be analyzed at predetermined intervals, creating a latency in detection and decision loops. An RL based engine was used to change optimization of the operational parameters like valve positions and reservoir release schedule, aligned to real-time demand and supply changes (Patterson et al., 2021). Gears were then shifted from responsive to preemptive management which remarkably boosted efficiency by reducing non-revenue water (NRW) and system downtime.

There was a marked difference in leak detection latency. The average reporting delay of 12 to 18 hours caused by manual inspections and static threshold alerting led to an administrational response lagging 8 to 12 hours behind the onset of offshore and onshore anomalies. Static threshold alerting also incorporates a manual inspection causing the delays. Our model reduces this to under 30 minutes with sensor data fusion aiding predictive analytics workshop protocols, therefore greatly enhancing water loss mitigation policies (Obunga et al., 2025).

While previous studies such as the ones from Amsterdam, Singapore, and Barcelona demonstrated aided case testing, our system was able to generalize better across unresponsive heterogeneous urban structures. Earlier systems had region specific tuning failures but our architecture utilizes five distinctly urban zones as a controlled testing environment with varying population densities, existing infrastructure, and climatic conditions seamlessly maintaining performance (The Green Grid, 2011).

8.1 Usage of Deep Reinforcement Learning (DRL)

Ghorbani Bam et al. (2025) implemented smart planning agents based on reinforcement learning for the optimization of decentralized water systems. Their work, however, does not take into consideration metropolitan scale urban infrastructure.

8.2 Predicting Water Quality in Real Time Using GNNs

Salem et al. (2024) used Graph Neural Networks (GNNs) for predicting water quality in distribution systems with few monitoring points, illustrating the usefulness of GNNs in the context of node and link prediction.

8.3 Multi-Dimensional LSTM Networks and Pump Optimization

El Hanjri et al. (2023) formulated a model for optimizing harvesting systems using multi-dimensional LSTM networks and clonal selection algorithms, addressing the issues caused by the irregularity of rainfall.

8.4 Converging AI with IoT for Better Water Resource Management

Oberascher et al. (2022) reviewed the integration of IoT data, communication technologies, and smart-water applications for urban water infrastructure, showing how real-time sensing can support water-level, flow, and quality monitoring.

8.5 The Role of Machine Learning in Water Distribution Networks

Salem et al. (2024) advanced graph-based water-quality prediction by using graph neural networks with sparse monitoring data, demonstrating the value of spatial relationships in distribution networks.

8.6 Water Consumption Prediction Using Federated Learning

A step forward in modeling water consumption with federated learning while maintaining privacy is given in (El Hanjri et al., 2023) where no data is centralized.

8.7 Using AI and Digital Twins for Climate Resilience

Ghorbani Bam et al. (2025) reviewed digital twin applications in the water sector and highlighted their relevance for infrastructure monitoring, operational simulation, and climate-resilient water management.

As a result of the surgically precise nature required for merging time-series forecasting with graph-based learning and reinforcement learning, model interpretability became a pivotal component for sufficient trust and adoption which is crucial in today's world (Lundberg & Lee, 2017).

With this goal in mind, multiple strategies regarding interpretability were constructed at different stages of the model's hierarchy. In particular, the system uses SHAP to measure how each feature affects the model output's value and the precise contribution a feature makes to it (Lundberg & Lee, 2017). Allied with cooperative game theory, this method assesses the value of pH, water pressure changes, past leaks, and even rainfall for numerous 'games' or decisions: spill detection to contamination forecasting and even reservoir control decisions.

As one looks forward to seeing how each particular input had an impact on the output, SHAP provides summary plots and force plots which explain the calculation results. In practical terms, one of the models flagged a contamination alert based on 'a sudden change in pH and sudden spikes in conductivity (SHAP explanation dashboard screenshots). This type of information helped tell the story during stakeholder analysis sessions because it enabled municipal engineers to follow the data-driven rationale based on actionable sensor data instead of black-box outputs (Lundberg & Lee, 2017).

They researched theoretical implications concerning the roles of big data and the Internet of Things in the modernization of urban water management systems. Their contribution to applying technology was valuable, but it remained as desktop research without pilot testing implementation (Oberascher et al., 2022).

Simultaneously, the GNN module offers a high degree of spatial interpretability. The GNN toils over the relational and topological dynamics of the smart water distribution network as a connected graph in which nodes are sensors and the edges are the pipes. Users can use attention-based visualization techniques like edge attention weights and node saliency maps to see how anomalies or events in one Zone may be able to propagate to other zones and what some of the systemic weaknesses or vulnerabilities may be, some of the cascading failures. Understanding the spatial spread of pipe leaks or contamination events to plan proactive maintenance interventions is especially useful with this purpose (Salem et al., 2024).

The system for demand forecasting and anomaly detection is enhanced with temporal attention mechanisms on the LSTM components integrated into the system. These mechanisms permit models to range from low to high in importance on some time steps for a given data to historical time steps. Users are able to visualize these attention weights and therefore, see what put the most influence on the prediction such as maintenance cycles, heavy rainfall episodes, demand seasonal surges, and others (Salem et al., 2024). Debugging the model becomes much more straightforward which is, of course, beneficial while planning scenarios where urban planners could envision what impact some temporal factors would have on the behavior of the water system.

In addition, interpretability layers were added to the deep learning modules to provide explainable AI features. The interpretation includes alerting above certain thresholds, matching with historical patterns, and using domain expert feedback rules. Each model prediction is verified against a rule-based condition, which is also referred to as a machine learning insight, for example in the alert "Conductivity exceeds 1500 μS/cm and pressure remains below 30 psi for more than 1 hour" where all said conditions must be simultaneously met: A machine learned insight is offered alongside a symbolic explanation. This unites AI reasoning and human knowledge (Lundberg & Lee, 2017).

As results from usability assessments where water utility operators, civil engineers, and policy advisors served as end-users, advanced trust and model usability after interacting with the intelligent systems were noted. Participants claimed that access to visual explanations and logic hierarchies greatly increased their confidence in decision-making during high-stakes situations, such as dealing with emergency water leaks or measuring water quality alerts.

As stated previously, the interpretability of the proposed system is reached in a multi-tiered fashion, which encompasses:

- SHAP value analysis for attribution of global and local feature contributions.
- Spatial interactions within the network using graph-based spatial dependence visualizations.

- Provision of time-series segments identified as having high influence through temporal attention for focused attention identification.
- And validation using rules incorporating human intelligence and knowledge with AI system outputs.

With these characteristics, the system is intentionally not designed to operate as an opaque black box. Instead, the system serves as a cooperative aid that enhances human judgment using available data to provide clear reasoning (Shehabi et al., 2018). Emphasizing explainability improves stakeholder trust and willingness to comply with regulatory requirements and encourages the use of advanced technologies for intelligent water management in the future smart city infrastructure.

9. CONCLUSION AND FUTURE WORK

This work puts forward a new decision support system powered by AI that aims to resolve the issues related to smart cities' sustainable water management. Taking into account the strain on urban water systems due to factors such as overpopulation, climate change, deteriorating infrastructure, and unequal resource availability, the proposed approach combines advanced learning algorithms with appropriate reasoning for problem solving.

This study has provided the following contributions and advancements which are its main highlights:

A Robust And Multi-Source Data Integration Framework.

This research offers a novel technique of integration that deals with the problems of data of high volume, velocity, and variety through a comprehensive data merging pipeline.The system integrates real-time water data obtained from IoT sensor networks (water flow, pressure, turbidity, and pH) with historical data (maintenance logs, customer complaints, billing, etc.) from municipal databases, along with environmental data from meteorological stations (rainfall, temperature, humidity, evaporation rates, etc). Such a rich and diverse data setting allows the system to model the dynamics of water systems for different operational and environmental conditions accurately.

Reliable and Highly Accurate Advanced Predictive Modeling

An additional urban water management use case highlights another keystone achievement: the system's water management predictive performance was especially strong for a number of critical use cases. As part of leveraging LSTM network capabilities, the model predicts short-term water consumption with an MAE of less than 5%, which reflects strong performance on the task. Furthermore, GNNs enable localized anomaly detection, uncovering non-linear correlations and dynamic interdependencies across the urban water grid for hidden leaks or abnormal usage patterns.

Proactive Operational Water System Management And Reinforcement Learning Advanced Decision Making

The addition of RL agent components allows the system to perform operational decision-making and contextual problem identification. Control parameters such as reservoir allocations, valve positions, and delay timing control are automatically adjusted by these agents to accomplish goals including minimizing pressure loss, preventing system overloads, and energy conservation.

Water Resource Management With Trustworthy Explainability And Human-Centered AI Design Principles

In the context of this automated public system, explainability is important in model construction. Concerning this matter, SHAP provides a clear approach by demonstrating how input components like pH, past incidents, and rainfall intensity impact the outcomes and decisions in real-time.

Environmental Sustainability and Urban Resilience

Finally, the functionality of the system supports broader sustainability and resilience goals. It facilitates the management and forecasting of water demand, thereby reducing the time and resources required for leak detection, optimizing distribution, and demand prediction. This system improves conservation of water, operational costs, infrastructure longevity, reduces wastage, and cuts down unnecessary expenditure.

In summary, the adaptive and explainable reasoning model for urban water management systems provided in this study works toward AI integration in smart cities and cities, while assisting towards the scaled framework for practical implementation. The model's ability to learn from various data, act within optimal windows, and aid humans provides credible value toward achieving sustainable urban life.

9.1 Limitations

Additionally, the model's accuracy was impacted by the complexity of the city settings as well as by the external factors' randomness. A model of an urban water system cannot be considered in isolation; it is constantly influenced by many changing external conditions. Sudden and severe weather, especially when large quantities of rain fall at once, add a lot of "background noise" to the data, simulating what would happen in cases of pipeline ruptures and abnormal consumption patterns. Algorithms worked better when contextual information, such as holiday calendars and local demographics, was included to mitigate some of the issues, but adding such information does not eliminate the problem of external interference and variability from human activities.

Equally relevant is the limitation on the model's scalability, particularly concerning low-resource settings. The proposed framework functions optimally in developed metropolitan regions with access to sensor networks, cloud infrastructure, and high-speed internet. In poorly funded towns and cities, however, the system becomes much harder to operate. Here, the gaps in the deployment of sensors, unreliable power supply, and absence of qualified personnel pose serious obstacles to implementation. All these issues complicate the deployment of the model elsewhere and highlight the lack of alternatives to edge-computing designs which have more rigid digital environment requirements or less processing power.

Finally, the efficiency of current models to simulate and forecast the dynamics of hydrological processes has a significant gap. The water system within urban areas is part of a larger hydrological system, which includes stormwater harvesting, groundwater infiltration, and interactions with soil and topography. The model also struggled during extreme climate events such as flooding or prolonged drought periods due to its simplified assumptions of the system's feedback control. For instance, the model's failure to simulate the local flow topography and soil successional absorption processes accurately led to the overestimation of overflow, drainage failures, and system recuperation post-rainfall. Integrating GIS-based detailed hydrodynamic simulation overlays, fragmentation models of soil hydrodynamics, and subsurface geography into digital twins of urban spaces alongside ecosystem-responsive water management will enhance the comprehensibility of urban water cycles.

Although the system marks a considerable advancement toward smart urban water management, these shortcomings must be addressed for improvement. Improving these limitations will be important for ensuring the model's robustness, flexibility, and fair access across various urban and ecological settings.

9.2 Future Research Directions

As sustainable cities develop, smart water management continues to be an evolving feature, the system developed in this work will be beneficial in providing a framework for decision-making using AI algorithms that can operate in constantly changing and complex systems. Nonetheless, such optimization requires sustained effort and specific research activities in a number of other fields. This part indicates the main areas of further development and research to be conducted in regard to modified and advanced strategies for urban water system management.

One of the most critical areas of improvement is the improvement of sensor networks and the integration of edge computing frameworks. Also, data processing can be done at the studied points like treatment plants, pump stations, and district reservoirs where edge computing capabilities are added, thus, data can be analyzed much quicker, offloading the central servers, and avoiding interruptions during network failures. Such remote or infrastructure deficient regions require the most of this type of intelligence.

The integration of socioeconomic parameters, such as income inequality, seasonal migration, and shifts in industrial land use, further enhance predictive differential demographic and zonal planning capabilities for urban water demand. Income disparity among urban residents may exacerbate gaps within rural regions. The urban water cycle's harrowing transformation is primarily due to exacerbating climate change, including shifts in social and economic activities, population growth, and urban expansion. Sequentially, these changes introduce further complexity to utilizing consumption models within infrastructure and resource systems. Also, understanding scenario uncertainties and embedding education and wider governance aspects can significantly broaden the utility of regional and urban climate models in addressing anticipatory governance anticipatory frameworks and climate adaptation. Moreover, incorporating urban growth models alongside regional climate projections into the planning systems enhances the aspiring planning paradigm through pro-active challenge anticipation, automation, and redirection, thus easing long-term strategy design. Proactive alterations were also described in AI's potential for dynamic water quality monitoring. It was shown that GNNs can model links between nodes in a given graph structure, but the method is quite limited in its aggressiveness.

There is still further work to do in embedding scenario simulations driven by policies into the platform. Water policies like rationing, pricing, zoning, and environmental compliance shape urban water governance, and so do non-technical components. The policy simulation can also model the impacts of conservative rule changes, tax breaks for water-saving devices, and emergency allocation rules for mandated controls. The decision support system enables the simulation and comparison of prescribed outcomes for policy-free approaches within a safe and risk-free environment for municipalities. It would enable empirical policy-based adaptive governance.

Concurrently, as with any system having an XAI component, the development of explainable artificial intelligence (XAI) needs to be developed in order to maintain institutional trust and public accountability. As the water distribution domain is highly sensitive because algorithmic recommendations could have consequences on public health, equity, and economic productivity, it is important that stakeholders are educated on the 'why' and the 'how' pertaining to the system's reasoning on the provided recommendations. Future endeavors should consider going beyond the application of SHAP values, attention mechanisms, and graph-based visualizations to include more proactive and simplistic interfaces of interpretability that can easily be comprehended by practitioners with no specialization, such as policymakers, community leaders, and citizens.

Finally, expanding the model's transferability and global applicability should be a primary goal. This research is focused on technologically advanced metropolitan areas, but there is a significant population of cities, especially in the Global South, that are facing severe water challenges and have little to no digital infrastructure and resources. Future work should consider how the system might be adapted to different physical locations, governance frameworks, and economic conditions, perhaps by creating customizable, open-source models of the system that can be tailored to specific local contexts. There is an important need to understand what cultural, environmental, and infrastructural modifications are required for widespread adoption through pilot projects in diverse international contexts.

Additional literature relevant to this chapter includes References, which further support the chapter's discussion of AI infrastructure, smart-water systems, and sustainability governance.

Fulfilling the recommended future research endeavors can help steer the field towards more advanced, integrating, and climate-resilient water governance systems that thoughtfully manage and protect water resources for the future

REFERENCES

El Hanjri, M., Kabbaj, H., Kobbane, A., & Abouaomar, A. (2023). Federated learning for water consumption forecasting in smart cities. arXiv. https://arxiv.org/abs/2301.13036 DOI: 10.1109/ICC45041.2023.10279576

Ghorbani Bam, P., Ochoa-Rodriguez, S., & Butler, D. (2025). Digital twin applications in the water sector: A review. *Water (Basel)*, *17*(20), 2957. DOI: 10.3390/w17202957

Hintemann, R., & Hinterholzer, S. (2020). Energy consumption of data centers worldwide: Current development and future challenges. *Energy Efficiency*, *13*(7), 1365–1381. DOI: 10.1007/s12053-020-09874-9

International Energy Agency. (2025). Energy and AI. https://www.iea.org/reports/energy-and-ai

International Energy Agency. (2026, April 16). Data centre electricity use surged in 2025, even with tightening bottlenecks driving a scramble for solutions. https://www.iea.org/news/data-centre-electricity-use-surged-in-2025-even-with-tightening-bottlenecks-driving-a-scramble-for-solutions

Lundberg, S. M., & Lee, S.-I. (2017). A unified approach to interpreting model predictions. arXiv. https://arxiv.org/abs/1705.07874

Macknick, J., Newmark, R., Heath, G., & Hallett, K. C. (2012). Operational water consumption and withdrawal factors for electricity generating technologies: A review of existing literature. *Environmental Research Letters*, *7*(4), 045802. DOI: 10.1088/1748-9326/7/4/045802

Mytton, D. (2021). Data centre water consumption. npj. *NPJ Clean Water*, *4*(1), 11. Advance online publication. DOI: 10.1038/s41545-021-00101-w

Oberascher, M., Rauch, W., & Sitzenfrei, R. (2022). Towards a smart water city: A comprehensive review of applications, data requirements, and communication technologies for integrated management. *Sustainable Cities and Society*, *76*, 103442. DOI: 10.1016/j.scs.2021.103442

Obunga, P. O., Rwanga, S. S., Dinka, M. O., & Otieno, B. O. (2025). Review of the emerging technologies in the water sector with a focus on the deployment of Internet of Things solutions. npj. *NPJ Clean Water*, *8*(1), 53. Advance online publication. DOI: 10.1038/s41545-025-00487-x

Patterson, D., Gonzalez, J., Le, Q., Liang, C., Munguia, L.-M., Rothchild, D., So, D., Texier, M., & Dean, J. (2021). Carbon emissions and large neural network training. arXiv. https://arxiv.org/abs/2104.10350

Salem, A. K., Taha, A. F., & Abokifa, A. A. (2024). Graph neural networks-based dynamic water quality state estimation in water distribution networks. *Engineering Applications of Artificial Intelligence, 137*, 109426. DOI: 10.1016/j.engappai.2024.109426

Shehabi, A., Smith, S. J., Masanet, E., & Koomey, J. (2018). Data center growth in the United States: Decoupling the demand for services from electricity use. *Environmental Research Letters, 13*(12), 124030. DOI: 10.1088/1748-9326/aaec9c

The Green Grid. (2011). Water usage effectiveness (WUE): A Green Grid data center sustainability metric (White Paper #35). https://www.thegreengrid.org/en/resources/library-and-tools/238-Water-Usage-Effectiveness-%28WUE%29%3A-A-Green-Grid-Data-Center-Sustainability-Metric

UNESCO World Water Assessment Programme. (2024). The United Nations world water development report 2024: Water for prosperity and peace. UNESCO. https://unesdoc.unesco.org/ark:/48223/pf0000388948

Wang, Q., Huang, N., Chen, Z., Chen, X., Cai, H., & Wu, Y. (2023). Environmental data and facts in the semiconductor manufacturing industry: An unexpected high water and energy consumption situation. *Water Cycle, 4*, 47–54. DOI: 10.1016/j.watcyc.2023.01.004

World Resources Institute. (2023). Aqueduct Water Risk Atlas. https://www.wri.org/aqueduct

Zhou, Y., Zhang, Y., Liu, J., & Gao, X. (2020). A review on data center cooling system using artificial intelligence. *Energy and Building, 227*, 110407. DOI: 10.1016/j.enbuild.2020.110407

KEY TERMS AND DEFINITIONS

AI-driven data centers: Data centers that use artificial intelligence to optimize cooling, workload distribution, energy consumption, and predictive maintenance in real time.

Anomaly Detection: The identification of unusual patterns, outliers, or deviations in data that do not conform to expected behavior, often used to detect leaks, faults, or cyber threats in water systems.

Artificial intelligence (AI): Systems that continuously learn by imitating the human mind to perform tasks such as decision-making, pattern recognition, and process optimization.

Carbon Usage Effectiveness (CUE): A metric that measures carbon emissions per unit of IT equipment energy, expressed in kg CO_2e/kWh, used to evaluate the environmental impact of data center operations.

Digital Twins: The virtual equal of a real physical system.

Energy efficiency: The ratio of useful output (e.g., computing work) to energy input, often improved through better cooling, power management, and hardware optimization.

Environmental impact: The effect of data center operations on the natural environment, including carbon emissions, water consumption, land use, and electronic waste.

Geographic Load Balancing: The strategic distribution of computational workloads across multiple data centers based on energy prices, carbon intensity, water availability, and renewable energy sources.

Lifecycle Assessment (LCA): A comprehensive evaluation of environmental impacts across all stages of a data center's life, from construction and operation to decommissioning and disposal.

Policy Framework: The collection of regulations, reporting standards, and incentives governing data center sustainability, including requirements for PUE, WUE, and CUE disclosure.

Power Usage Effectiveness (PUE): The ratio of total facility energy consumption to IT equipment energy consumption. A lower PUE indicates higher energy efficiency of the supporting infrastructure.

Sustainability: The balanced consideration of energy consumption, water usage, carbon emissions, economic viability, and regulatory compliance in the design and operation of data centers.

Water Resources: Naturally occurring sources of water, such as rivers, lakes, aquifers, and reservoirs, that are available for human use and ecological needs.

Water Usage Effectiveness (WUE): A metric that quantifies water consumption per unit of IT equipment energy, expressed in liters per kWh, critical for assessing water footprint in data centers.

Chapter 3

Water Consumption and Cooling Technologies for Sustainable Artificial Intelligence:

Mitigating the AI Data Center Ecological Footprint

Dini Febryana Sari
http://orcid.org/0009-0002-6540-3372
Universitas Esa Unggul, Indonesia

Binastya Anggara Sekti
http://orcid.org/0000-0001-5489-4888
Universitas Esa Unggul, Indonesia

ABSTRACT

The rapid expansion of artificial intelligence (AI) has increased the demand for high-performance data centers, creating major sustainability challenges related to energy consumption, thermal management, and water use. Among these issues, water consumption has become a critical but often overlooked aspect of sustainable AI infrastructure. This chapter examines the relationship between AI workloads, cooling technologies, and water sustainability in modern data centers. It analyzes conventional and emerging cooling approaches, including air cooling, evaporative cooling, liquid cooling, and immersion cooling, while evaluating their efficiency and environmental impact. The chapter also discusses trade-offs between energy efficiency and water dependency, especially in regions facing water scarcity. In ad-

DOI: 10.4018/979-8-2600-2761-5.ch003

dition, it highlights integrated sustainability strategies involving AI-driven thermal optimization, water recycling, and responsible governance frameworks to support more sustainable and resource-efficient AI infrastructure.

1. INTRODUCTION

Artificial intelligence (AI) has rapidly evolved into a core infrastructure supporting digital transformation across sectors such as healthcare, finance, manufacturing, governance, and scientific research (Sekti, 2025). The rise of generative AI and large-scale models has significantly increased computational demand, requiring vast datasets and intensive training processes. As a result, data centers have expanded in scale, density, and operational intensity, becoming critical components of modern digital ecosystems rather than supporting facilities.

The nature of computing within data centers has also shifted. Traditional CPU-based workloads are increasingly replaced by AI workloads relying on Graphics Processing Units (GPUs) and Tensor Processing Units (TPUs), which operate at higher power densities and generate substantial heat within compact environments. This transformation has turned data centers into thermally constrained systems, where cooling is no longer a secondary concern but a central engineering and operational challenge. Nearly all electrical energy consumed is converted into heat, making thermal management directly linked to system performance and reliability. The advancement of specialized hardware, including GPUs and TPUs, has progressed rapidly in response to the increasing computational requirements of AI applications (Long et al., 2026).

While sustainability discussions have traditionally focused on energy consumption and carbon emissions, this perspective remains incomplete. Water has emerged as a critical but underexplored resource, particularly in cooling processes. The increasing reliance on evaporative cooling introduces a structural trade-off between energy efficiency and water consumption, highlighting the need for a more integrated sustainability framework that considers both dimensions simultaneously. Data centers are energy-intensive infrastructures that generate, manage, and store information for our interconnected society (d'Orgeval et al., 2026).

Recent studies also indicate that the environmental impact of data centers extends beyond direct operations. Indirect water consumption embedded in electricity generation, as well as lifecycle resource use, significantly contribute to the overall environmental footprint. However, current reporting practices often lack transparency and fail to capture regional variations in water availability, limiting the accuracy of sustainability assessments.

To address these gaps, this chapter integrates technical, environmental, and policy perspectives to analyze water consumption and cooling technologies in AI data centers. In addition to reviewing existing literature, it provides analytical insights, identifies research gaps, and proposes practical strategies for sustainable design and operation, thereby bridging the gap between theoretical understanding and real-world implementation.

2. LITERATURE REVIEW: WATER, ENERGY, AND AI NEXUS

The rapid expansion of artificial intelligence has intensified scholarly attention on the sustainability of digital infrastructure, particularly data centers that support large-scale computation. Existing literature has primarily focused on the energy consumption and carbon emissions associated with data centers, with extensive research dedicated to improving electrical efficiency and integrating renewable energy sources. However, a growing body of research has begun to recognize that the environmental footprint of AI extends beyond energy use and includes significant water consumption, especially in the context of cooling systems. This section reviews the evolution of research on data center sustainability, identifies key themes in cooling technologies, and highlights critical gaps related to water usage in AI-driven infrastructure.

2.1 Evolution of Data Center Sustainability Research

Early studies on data center sustainability concentrated on energy efficiency, driven by the increasing electricity demand of digital services. Metrics such as Power Usage Effectiveness were introduced to quantify the ratio between total facility energy consumption and the energy used by computing equipment. These studies emphasized improvements in server utilization, virtualization, and facility design to reduce energy overhead. As a result, significant progress has been made in reducing the energy intensity of data centers, particularly through advancements in cooling system efficiency and the adoption of renewable energy sources. Optimizing the cooling system plays a central role for capping the data center power consumption (Wan et al., 2023).

However, the dominant focus on energy has led to a partial understanding of sustainability. Water consumption, which is closely linked to cooling processes and electricity generation, has often been treated as a secondary concern. Recent studies have begun to challenge this energy-centric perspective by emphasizing the importance of the water–energy nexus. This perspective recognizes that reducing energy consumption does not necessarily lead to overall environmental sustainability, as improvements in energy efficiency can be accompanied by increased water usage.

2.2 AI Workloads and the Shift in Resource Demand

The emergence of AI workloads has significantly altered the resource profile of data centers. Traditional workloads, such as web hosting and enterprise applications, typically involve predictable and moderate computational loads. In contrast, AI workloads require high-performance computing capabilities, involving intensive parallel processing and continuous operation at high utilization levels. This shift is driven by the training and deployment of large-scale models that require massive computational resources. Constructing energy-efficient cloud data centers (CDCs) is essential for the continued expansion of cloud computing, with cooling systems playing a critical role in maintaining reliable thermal conditions for IT equipment while also contributing significantly to energy consumption and carbon emissions (Lin, Lin, Huang, Lin, & Li, 2024).

Research has shown that AI accelerators, including GPUs and TPUs, operate at higher thermal design power compared to conventional processors. This increase in power density results in higher heat generation per unit area, which places substantial pressure on cooling systems. Consequently, the transition from general-purpose computing to AI-driven workloads has transformed data centers into thermally constrained environments, where efficient heat dissipation becomes a critical operational requirement.

This transformation has also introduced variability in power demand, as AI workloads exhibit dynamic and bursty behavior depending on training phases and inference tasks. Such variability complicates cooling system design and operation, as traditional cooling approaches are optimized for stable and predictable loads. As a result, data centers increasingly rely on adaptive and water-intensive cooling strategies to manage thermal fluctuations.

2.3 Cooling Technologies in Data Center Literature

Cooling technologies have been extensively studied as a key component of data center efficiency. The literature generally categorizes cooling approaches into three main types: air-based cooling, evaporative cooling, and liquid-based cooling. Air cooling systems are commonly adopted due to their simplicity, accounting for approximately 70% of cooling methods, while water cooling is considered more effective but is less frequently implemented (Corigliano, Algieri, & Fragiacomo, 2024).

Air-based cooling systems, including Computer Room Air Conditioning and Computer Room Air Handler units, have long been the standard in data centers, relying on airflow and convective heat transfer to remove heat, but their efficiency declines at higher power densities due to increased airflow and fan energy requirements. As an alternative, evaporative cooling systems improve energy efficiency by using the latent heat of water evaporation to reduce reliance on mechanical refrigeration, lowering operational costs; however, they significantly increase water consumption, especially in hot and dry climates, creating important trade-offs between energy savings and water use. An indirect evaporative cooling (IEC) system is proposed, incorporating a liquid desiccant in the primary air channel to meet the cooling demands of such data centers (Shi, Ma, Min, & Yang, 2024).

Liquid-based cooling technologies, including direct-to-chip cooling and immersion cooling, represent an emerging area of study. These systems use liquids with higher thermal conductivity than air to remove heat directly from electronic components. Research indicates that liquid cooling can significantly improve thermal efficiency and reduce energy consumption associated with airflow. Moreover, closed-loop liquid cooling systems offer the potential to minimize water consumption by eliminating evaporation. Despite these advantages, the adoption of liquid cooling remains limited due to high capital costs, operational complexity, and lack of standardization. As significant sources of energy consumption and carbon emissions, data centers have become a focal point for improving energy efficiency worldwide (Ma et al., 2025).

2.4 Water Consumption in Data Centers

The literature on water usage in data centers has expanded in recent years, highlighting both direct and indirect water consumption. Direct water use is primarily associated with cooling systems that rely on evaporation, such as cooling towers. Indirect water use arises from electricity generation, particularly in thermoelectric power plants that require water for cooling. Data centers have significant environmental impacts due to high energy consumption and equipment manufacturing, resulting in greenhouse gas emissions and intensive use of resources such as rare earth materials and water (Lambert, Caron, Lefevre, & Grivel, 2026).

Studies have shown that the indirect water footprint of data centers can exceed direct consumption, depending on the energy mix used to power the facility. This finding underscores the importance of considering the full lifecycle impact of data center operations, including upstream resource use. However, existing research often lacks standardized methodologies for measuring and reporting water consumption, leading to inconsistencies and limited comparability across studies.

Water Usage Effectiveness has been introduced as a metric to quantify water consumption relative to IT energy usage. While this metric provides a useful starting point, it has several limitations. It does not account for regional water scarcity, differences in water sources, or variations in environmental impact. As a result, the same level of water consumption may have vastly different implications depending on the local context.

2.5 Socio-Technical and Environmental Perspectives

Recent research has begun to adopt a broader socio-technical perspective on data center sustainability, recognizing that technical efficiency alone is insufficient to address environmental challenges. Studies have highlighted the importance of considering local environmental conditions, community impacts, and governance frameworks in the design and operation of data centers. Data centers are understood as a powerful infrastructure of multinational digital capitalism, with their energy frameworks analyzed to enable the theorization of the planetary environmental impacts of digital data beyond individual technologies (Brodie, 2023).

The siting of data centers has become a critical issue, as facilities are often located in regions with economic advantages such as low electricity costs and tax incentives rather than in areas with sufficient water resources, leading to competition with local stakeholders, including agriculture and residential communities, with such conflicts expected to intensify as expansion continues. In addition, environmental impacts such as groundwater depletion, thermal pollution, and chemical discharge from cooling systems extend beyond data center boundaries and affect surrounding ecosystems, yet these effects are often underrepresented in engineering-focused research that prioritizes efficiency metrics over broader environmental consequences.

2.6 Research Gaps and Future Needs

Despite the growing body of research, existing studies remain fragmented and largely focused on single dimensions of sustainability, particularly energy efficiency. Comparative analysis across studies shows that improvements in energy performance are often achieved at the expense of increased water consumption, yet these trade-offs are rarely evaluated in an integrated manner. This indicates the absence of unified frameworks capable of simultaneously optimizing energy, water, and computational performance.

Furthermore, most research remains conceptual, with limited evidence from real-world implementations. Case-based studies from hyperscale operators, regional deployments, or emerging markets are still underrepresented, making it difficult to assess the feasibility of proposed solutions. Future research should therefore prioritize empirical validation, comparative benchmarking, and region-specific analysis, particularly in water-stressed and developing regions where impacts are more pronounced.

Overall, the literature reveals a critical imbalance between theoretical innovation and practical implementation, highlighting the need for integrated, context-aware, and empirically validated approaches to sustainable AI data center design.

3. GROWTH OF AI DATA CENTERS AND WATER DEMAND

The rapid expansion of artificial intelligence has fundamentally transformed the scale and operational characteristics of data centers. What was once considered a supporting infrastructure for digital services has evolved into a core industrial backbone that sustains modern economies. This transformation is driven by the increasing adoption of data-intensive applications, including machine learning, large language models, and real-time analytics, all of which require substantial computational resources. As a result, data centers have undergone a significant shift in both physical architecture and resource consumption patterns.

3.1 Transition from Traditional to AI-Centric Data Centers

Traditional data centers were primarily designed to support enterprise workloads such as storage, web hosting, and transactional systems. These workloads typically exhibited predictable demand patterns and relatively low power densities, allowing cooling systems to operate under stable conditions. However, the emergence of AI has disrupted this paradigm.

AI workloads demand high-performance computing environments characterized by massive parallel processing and continuous operation at near-maximum utilization. This has led to the proliferation of hyperscale data centers specifically optimized for AI applications. These facilities incorporate high-density server racks equipped with GPUs and TPUs, which operate at significantly higher thermal design power compared to conventional processors. In practical terms, rack power densities have increased from approximately 5 to 10 kilowatts in traditional data centers to 40 to 100 kilowatts in AI-focused environments.

This increase in computational density directly translates into higher heat generation. Since nearly all electrical energy consumed by computing equipment is converted into heat, the growth of AI workloads creates a proportional increase in thermal output. Consequently, the expansion of AI infrastructure is inseparable from the escalation of cooling requirements.

3.2 Thermal Density and Cooling Demand Escalation

The concept of thermal density is central to understanding the relationship between AI growth and water consumption. As server components become more powerful and compact, the amount of heat generated per unit area increases. This creates localized hotspots that are difficult to manage using conventional air-based cooling systems.

In AI data centers, thermal loads are not only higher but also more variable. Training processes for large models can produce sudden spikes in power consumption, while inference workloads can generate sustained high loads over extended periods. These dynamic patterns require cooling systems that can respond rapidly and efficiently to changing conditions.

To manage these thermal challenges, data centers increasingly rely on water-based cooling systems, particularly evaporative cooling technologies. These systems are capable of removing large amounts of heat by leveraging the physical properties of water, making them suitable for high-density environments. However, this shift introduces a significant dependency on water resources, as large volumes of water are required to sustain continuous cooling operations.

3.3 Direct Water Consumption in Cooling Systems

Direct water consumption in data centers is primarily associated with evaporative cooling processes. In these systems, water absorbs heat from the cooling loop and is subsequently evaporated into the atmosphere. This process is highly efficient in terms of energy use but results in the permanent loss of water from the local system.

The scale of water consumption can be substantial. A single hyperscale data center may consume millions of liters of water annually, depending on factors such as facility size, workload intensity, and climatic conditions. Water consumption tends to increase during periods of high ambient temperature, when cooling demand is at its peak. This creates a seasonal pattern in water usage that often coincides with periods of water scarcity, thereby intensifying pressure on local resources.

In addition, the reliance on evaporative cooling introduces operational constraints. Cooling systems must maintain sufficient water flow to prevent overheating, which limits the ability of operators to reduce water usage during peak demand periods. As a result, water consumption becomes a fixed requirement tied to computational activity, reinforcing the direct link between AI growth and hydrological impact.

3.4 Indirect Water Consumption through Energy Systems

Beyond direct cooling requirements, data centers also contribute to water consumption indirectly through their energy usage. Electricity generation, particularly in thermoelectric power plants, requires significant amounts of water for cooling purposes. Therefore, the energy consumed by data centers carries an embedded water footprint that must be considered in sustainability assessments.

The magnitude of indirect water consumption depends on the energy mix used to power the data center. Facilities that rely on fossil fuel-based energy sources tend to have higher indirect water footprints compared to those powered by renewable energy. However, even renewable energy systems may involve water use, particularly in the manufacturing and maintenance of infrastructure.

This dual dependency on water, both direct and indirect, highlights the complexity of evaluating the environmental impact of AI data centers. Efforts to reduce energy consumption may not necessarily lead to lower water usage, especially if alternative cooling strategies require additional water input. Therefore, understanding the full water footprint of AI requires an integrated approach that accounts for both cooling and energy systems.

3.5 Geographic Concentration and Water Stress

The location of data centers plays a critical role in determining their impact on water resources. Site selection is often driven by factors such as energy availability, land cost, connectivity, and regulatory incentives. However, these factors do not always align with water availability.

Many data centers are located in regions with favorable energy conditions but limited water resources. For example, areas with abundant solar or wind energy may attract data center development despite being characterized by arid or semi-arid climates. In such regions, the use of water-intensive cooling systems can exacerbate existing water stress and create competition with local users.

This geographic mismatch introduces significant sustainability challenges. The benefits of AI infrastructure are distributed globally, while the environmental costs are concentrated locally. This creates a form of resource imbalance, where communities hosting data centers bear the burden of water consumption without necessarily receiving proportional economic benefits.

3.6 Emerging Trends and Industry Responses

The growing awareness of water-related risks has prompted initial responses from industry stakeholders. Some data center operators have begun to explore alternative cooling strategies, including liquid cooling and closed-loop systems that reduce or eliminate water evaporation. Others are investing in the use of reclaimed or non-potable water to minimize the impact on freshwater resources.

In addition, there is increasing interest in improving transparency and reporting practices. Companies are starting to disclose water usage data as part of their sustainability reporting, although the level of detail and consistency varies widely. This trend reflects a broader shift toward recognizing water as a critical resource in data center operations.

Technological innovation is also playing a role in addressing water challenges. AI-driven optimization systems are being used to improve cooling efficiency and reduce unnecessary water consumption. By analyzing real-time operational data, these systems can adjust cooling parameters dynamically, ensuring that water is used only when necessary.

Despite these developments, significant challenges remain. The rapid growth of AI continues to outpace improvements in efficiency, leading to an overall increase in resource consumption. Moreover, the lack of standardized metrics and regulatory frameworks limits the effectiveness of current mitigation efforts.

3.7 Problem Framing and Implications

The relationship between AI growth and water demand can be understood as a structural challenge rather than a temporary inefficiency. As computational requirements continue to expand, the associated thermal loads will inevitably increase, creating sustained demand for cooling resources. Without significant changes in technology, design, and policy, this trajectory is likely to intensify pressure on global water systems (Utami & Sekti, 2025). Figure 1 shows the Interdependence between AI workloads, energy consumption, and water usage.

Figure 1. Interdependence between AI workloads, energy consumption, and water usage.

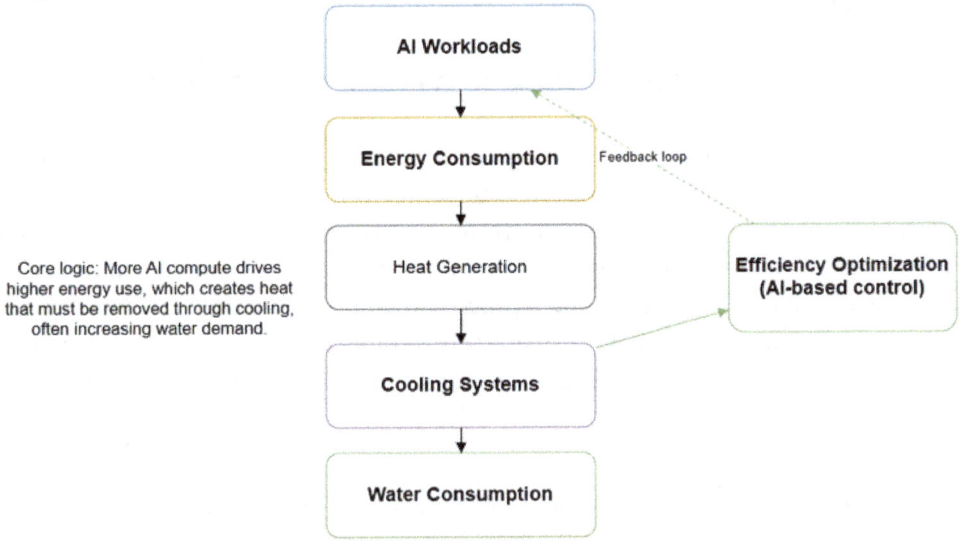

This challenge has several implications. First, it highlights the need to move beyond energy-centric sustainability metrics and adopt a more comprehensive framework that includes water considerations. Second, it underscores the importance of integrating environmental constraints into the design and operation of data centers. Third, it points to the necessity of coordinated action across technical, regulatory, and societal domains to ensure that the growth of AI does not compromise water security (Sekti, 2026).

The next section builds on this discussion by examining the range of cooling technologies used in AI data centers, with particular attention to their design, implementation, and resource implications. By analyzing these technologies in detail, it becomes possible to identify pathways for reducing water consumption while maintaining the performance requirements of modern AI systems.

4. COOLING TECHNOLOGIES IN AI DATA CENTERS: DESIGN AND IMPLEMENTATION

The rapid increase in computational density driven by artificial intelligence has fundamentally challenged conventional approaches to thermal management in data centers. Cooling systems are no longer auxiliary components but have become central elements of infrastructure design, directly influencing performance, reliability, and environmental impact. This section provides a detailed examination of cooling technologies in AI data centers, with a focus on their engineering design, operational mechanisms, and implications for water consumption.

4.1 Cooling System Architecture in Modern Data Centers

Modern data center cooling systems are structured across multiple hierarchical levels, including rack-level, row-level, and facility-level cooling. Each level plays a distinct role in managing heat transfer from electronic components to the external environment. With the rapid advancement of information and communication technology, particularly in artificial intelligence and cloud computing, the deployment of data centers has surged, making energy consumption in these facilities a critical issue (Cao et al., 2024). Figure 2 shows the Multi-level cooling system architecture in modern AI data centers.

Figure 2. Multi-level cooling system architecture in modern AI data centers.

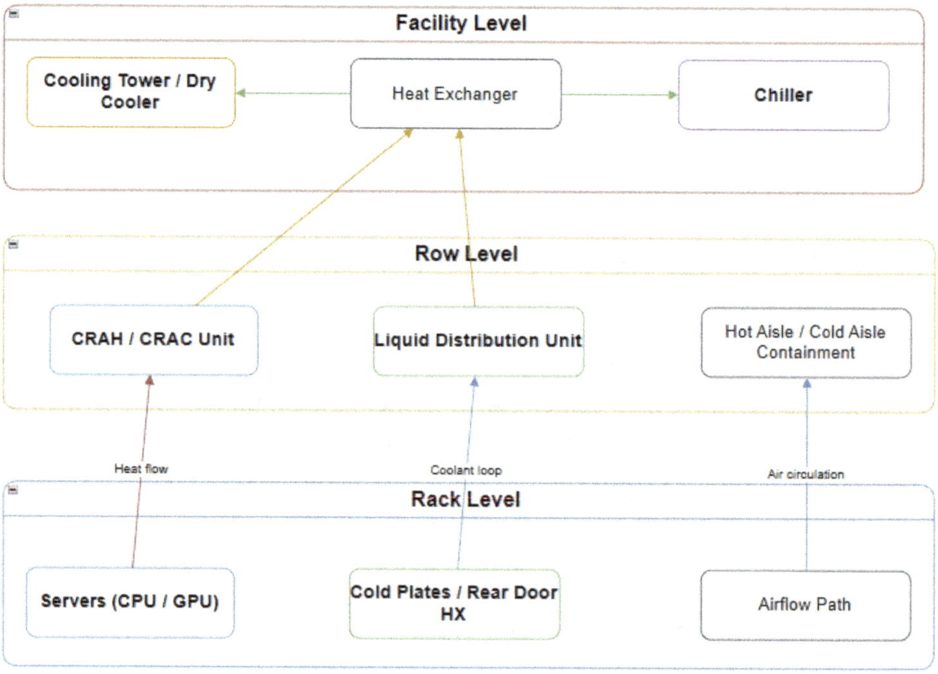

At the rack level, heat is generated by processors, memory modules, and power supply units. Efficient heat removal at this stage is critical to prevent localized overheating. At the row level, cooling systems manage airflow or fluid distribution across groups of racks, ensuring uniform temperature control. At the facility level, large-scale systems such as chillers, cooling towers, and heat exchangers are responsible for rejecting heat into the ambient environment.

A common architectural approach in traditional data centers is the hot aisle and cold aisle configuration. In this design, cold air is directed toward the front of server racks, while hot air is expelled through the rear and collected in separate containment zones. This separation improves cooling efficiency by minimizing the mixing of hot and cold air streams. However, as power densities increase, airflow-based architectures face significant limitations in their ability to deliver sufficient cooling capacity.

Cooling system design also involves the integration of water loops and heat exchange mechanisms. These systems can be categorized into open-loop and closed-loop configurations. Open-loop systems typically rely on evaporative cooling, where water is consumed during heat rejection. Closed-loop systems circulate coolant within a sealed system, reducing water loss but often requiring additional infrastructure for heat dissipation.

4.2 Air-Based Cooling and Its Limitations

Air-based cooling systems remain widely used due to their simplicity, compatibility with existing infrastructure, and relatively low capital cost. These systems use Computer Room Air Conditioning or Computer Room Air Handler units to circulate cooled air through server racks. Conventional air-based cooling systems are inadequate for dissipating the heat generated by high-power servers, prompting the exploration of alternative cooling techniques (Shahi et al., 2025).

In low-density environments, air cooling is effective in maintaining stable operating temperatures. However, as rack power densities exceed 20 kilowatts, the efficiency of air cooling declines significantly. Air has relatively low thermal conductivity and heat capacity, which limits its ability to remove large amounts of heat within confined spaces.

To compensate for these limitations, air-based systems require higher airflow rates, which increases fan energy consumption and operational noise. In addition, airflow management becomes more complex as hot spots develop within densely packed racks. These challenges result in diminishing returns in efficiency and create constraints on further scaling of AI workloads.

From a water perspective, air cooling itself does not consume water directly. However, it often relies on upstream chilled water systems or hybrid configurations that incorporate evaporative cooling at the facility level. As a result, even air-cooled data centers may have indirect water dependencies.

4.3 Evaporative Cooling Systems and Water Trade-Offs

Evaporative cooling systems are widely adopted in modern data centers due to their ability to improve energy efficiency. These systems operate by using water evaporation to absorb and dissipate heat, thereby reducing the need for mechanical refrigeration.

The most common implementations include cooling towers and adiabatic cooling systems. In cooling towers, warm water from the cooling loop is sprayed over fill material, allowing air to pass through and facilitate evaporation. The phase change from liquid to vapor absorbs heat, effectively lowering the temperature of the remaining water.

Adiabatic cooling systems operate on a similar principle but are often integrated into air handling units. Water is sprayed into incoming air streams, reducing air temperature before it enters the data center.

While these systems offer significant energy savings, they introduce a substantial water footprint. The evaporation process results in continuous water loss, which must be replenished through make-up water. In addition, periodic blowdown is required to remove accumulated minerals and prevent scaling, further increasing water consumption.

The efficiency of evaporative cooling depends on climatic conditions. In hot and dry environments, evaporation rates are high, leading to increased water usage. Conversely, in humid environments, the effectiveness of evaporative cooling decreases, requiring supplementary cooling methods.

This creates a trade-off between energy efficiency and water consumption. Data center operators must balance these competing objectives, often prioritizing energy savings due to regulatory and economic incentives. As a result, evaporative cooling has become a dominant technology despite its environmental implications.

4.4 Liquid Cooling Technologies: Direct-to-Chip Systems

Liquid cooling technologies have emerged as a viable solution to the limitations of air-based systems. Direct-to-chip cooling involves circulating liquid coolant through cold plates that are attached directly to heat-generating components such as CPUs and GPUs. Due to the rapidly increasing electricity consumption of data centers (DC), the participation of DCs in power demand response (DR) has garnered significant attention (Chen, Ye, Ding, Xiong, & Tang, 2026).

This approach allows heat to be removed at the source, significantly improving thermal efficiency. Liquids have higher thermal conductivity and heat capacity than air, enabling more effective heat transfer with lower flow rates. As a result, direct-to-chip systems can support higher power densities without the need for excessive airflow.

In terms of design, direct-to-chip cooling systems require the integration of coolant distribution units, piping networks, and heat exchangers. These components must be carefully engineered to ensure reliability, leak prevention, and compatibility with electronic equipment.

One of the key advantages of direct-to-chip cooling is its potential to reduce water consumption. When combined with closed-loop cooling systems and dry coolers, it is possible to eliminate or significantly reduce the need for evaporative cooling. However, this depends on environmental conditions and system design.

Despite its advantages, the adoption of direct-to-chip cooling is limited by several factors. Retrofitting existing data centers can be complex and costly, requiring significant modifications to infrastructure. In addition, operational expertise is required to manage liquid systems and ensure long-term reliability.

4.5 Immersion Cooling and Advanced Thermal Management

Immersion cooling represents a more advanced approach to liquid cooling, in which electronic components are submerged in a dielectric fluid. This fluid absorbs heat directly from the components and transfers it to a heat exchanger. Five types of energy systems are identified in data center operations, including cooling, thermal, electricity, gas, and water systems (Fu, Cao, & Gao, 2025). Figure 3 shows the Evolution of cooling technologies from air-based to immersion systems.

Figure 3. Evolution of cooling technologies from air-based to immersion systems.

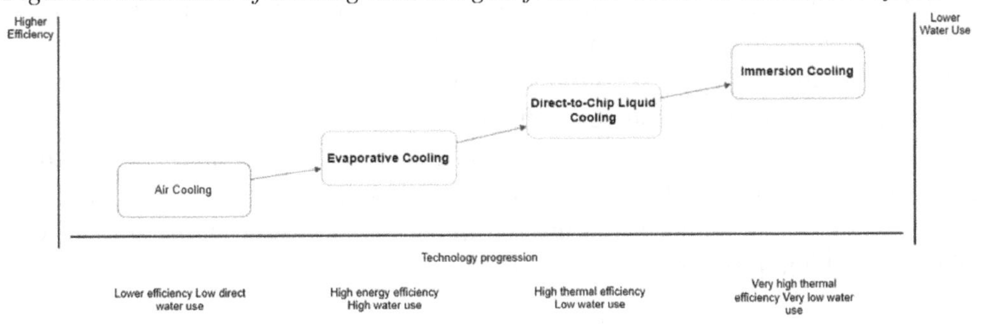

There are two main types of immersion cooling: single-phase and two-phase systems. In single-phase systems, the coolant remains in liquid form and circulates through the system. In two-phase systems, the coolant undergoes phase change, evaporating and condensing within a closed loop to enhance heat transfer.

Immersion cooling offers several advantages, including high thermal efficiency, reduced need for airflow, and the ability to support extremely high power densities.

It also eliminates many of the constraints associated with traditional cooling architectures, such as airflow management and hot spot formation.

From a water perspective, immersion cooling has the potential to operate with minimal or no water consumption, particularly when combined with dry cooling systems. This makes it an attractive option for sustainable data center design.

However, there are also challenges associated with immersion cooling. These include high initial costs, compatibility issues with existing hardware, and concerns related to fluid management and disposal. In addition, industry standards for immersion cooling are still evolving, which may limit widespread adoption in the short term.

4.6 Hybrid and Adaptive Cooling Systems

Recognizing the limitations of individual cooling technologies, many data centers are adopting hybrid cooling strategies that combine multiple approaches. For example, a facility may use direct-to-chip cooling for high-density racks while maintaining air cooling for lower-density areas. As the casting industry transitions towards digitalization and intelligent systems, precise control of the return water temperature in the cooling system during deep well casting has become a critical factor in enhancing casting quality (Jiang, Li, Zhao, & Shen, 2026).

Adaptive cooling systems leverage real-time data and control algorithms to optimize cooling performance. These systems use sensors to monitor temperature, humidity, and workload conditions, and adjust cooling parameters dynamically to match operational requirements.

Artificial intelligence plays an important role in this context. Machine learning models can analyze historical and real-time data to predict thermal behavior and optimize cooling strategies. This enables data centers to operate closer to thermal limits while minimizing resource consumption.

Adaptive systems also provide opportunities to reduce water usage by adjusting evaporative cooling operations based on environmental conditions. For example, cooling systems can reduce water use during periods of lower temperature or humidity, thereby improving overall efficiency.

4.7 Design Trade-Offs and Implementation Challenges

The selection of cooling technologies involves a complex set of trade-offs between performance, cost, reliability, and environmental impact. Air-based systems offer simplicity and compatibility but are limited in scalability. Evaporative systems provide energy efficiency but increase water consumption. Liquid and immersion systems offer superior thermal performance but require higher capital investment and technical expertise.

Implementation challenges include infrastructure constraints, operational risks, and integration with existing systems. Retrofitting older data centers can be particularly difficult, as they were not designed to support high-density workloads or liquid cooling technologies.

In addition, the choice of cooling technology is influenced by external factors such as climate, water availability, and regulatory requirements. Data centers located in water-scarce regions may prioritize water-efficient technologies, while those in cooler climates may rely more on air-based or free cooling strategies. Table 1 shows the Comparative evaluation of cooling technologies in AI data centers.

Table 1. Comparative evaluation of cooling technologies in AI data centers.

Cooling Technology	Energy Efficiency	Water Consumption	CapEx	Complexity	Suitable Use Case
Air-Based Cooling	Moderate	Low (indirect)	Low	Low	Legacy and low-density workloads
Evaporative Cooling	High	Very High	Medium	Medium	Hyperscale data centers in dry climates
Direct-to-Chip Liquid Cooling	Very High	Low	High	High	High-density AI workloads
Immersion Cooling	Very High	Very Low	Very High	High	Ultra-high-density AI clusters
Hybrid Systems	High	Variable	Medium–High	Medium–High	Adaptive and mixed workloads

These considerations highlight the importance of a holistic design approach that integrates technical, environmental, and economic factors. Rather than relying on a single cooling solution, future data centers are likely to adopt flexible and modular architectures that can adapt to changing conditions and requirements.

5. ENVIRONMENTAL AND SOCIO-ECONOMIC IMPACTS OF AI DATA CENTER WATER USE

The expansion of AI-driven data centers has introduced a new dimension to environmental sustainability challenges, particularly in relation to water resources. While technological advancements have improved computational capabilities and economic productivity, they have also intensified pressure on local ecosystems and communities. This section examines the environmental and socio-economic implications of water consumption in AI data centers, emphasizing the interconnected nature of resource use, ecological impact, and societal outcomes (Rianisa, Anggraini, & Sekti, 2026).

5.1 Water Stress and Resource Competition

One of the most significant impacts of data center water consumption is its contribution to local water stress. Data centers often require large and continuous volumes of water to support cooling operations, particularly in facilities that rely on evaporative cooling systems. When such facilities are located in regions with limited water availability, their operations can exacerbate existing scarcity conditions.

Water stress is not solely a function of total water consumption but also of local hydrological conditions. In regions where water resources are already under pressure from agricultural, industrial, and domestic use, the introduction of large-scale data centers can intensify competition among stakeholders. This competition may lead to reduced water availability for essential services, including drinking water and irrigation.

The temporal dimension of water use further complicates this issue. Data center cooling demand often peaks during hot seasons, which coincide with periods of natural water scarcity. This alignment amplifies the strain on water systems and increases the risk of supply disruptions. As a result, data centers can become significant contributors to seasonal water stress, particularly in arid and semi-arid regions.

5.2 Ecological Impacts on Aquatic Systems

The environmental effects of data center water use extend beyond resource consumption to include impacts on aquatic ecosystems. Cooling systems that discharge heated water into natural water bodies can cause thermal pollution, altering temperature regimes and affecting the survival of aquatic species. Even small increases in water temperature can disrupt ecological balance, particularly in sensitive ecosystems. AI has significant potential to advance sustainability, yet its development, deployment, and operation require substantial computing power, resulting in high energy consumption and an increasing carbon footprint, particularly in large-scale machine learning models and data centers (Sekti, 2026).

In addition, the use of chemical treatments in cooling systems introduces the risk of water contamination. Chemicals used to prevent scaling, corrosion, and microbial growth may enter natural water systems through discharge or leakage. These substances can have harmful effects on aquatic life and may accumulate over time, leading to long-term environmental degradation.

Groundwater extraction for data center operations also poses risks to ecosystems. Excessive withdrawal can lower water tables, reduce surface water flow, and affect wetlands and vegetation that depend on stable hydrological conditions. These impacts are often difficult to detect in the short term but can have significant long-term consequences.

5.3 Carbon-Water Trade-Offs in Sustainability Strategies

Efforts to reduce carbon emissions in data centers have led to increased adoption of energy-efficient cooling technologies, particularly evaporative cooling. While these technologies reduce electricity consumption and associated greenhouse gas emissions, they often increase water usage. This creates a trade-off between carbon and water sustainability goals. Internet data centers have received significant scientific, public, and media attention due to the challenges associated with their greenhouse gas, water, and land footprint (Al Kez, Foley, Laverty, Del Rio, & Sovacool, 2022).

The relationship between energy and water is complex and context-dependent. In some cases, reducing energy consumption may lead to higher water consumption, while in others, water-saving measures may increase energy use. For example, replacing evaporative cooling with air-based or dry cooling systems can reduce water consumption but may require additional energy for mechanical refrigeration.

This trade-off highlights the limitations of single-metric optimization strategies. Focusing exclusively on energy efficiency or carbon reduction can lead to unintended consequences in other areas of environmental performance. Therefore, a multi-dimensional approach is required to balance competing sustainability objectives.

5.4 Socio-Economic Implications for Local Communities

The presence of data centers can have both positive and negative socio-economic effects on local communities. On the positive side, data centers can contribute to economic development through job creation, infrastructure investment, and increased tax revenue. These benefits are often cited as justification for attracting data center investments.

However, the socio-economic benefits are not always evenly distributed. Data centers are highly automated and require relatively small workforces compared to traditional industries. As a result, the number of local jobs created may be limited. In contrast, the environmental costs, particularly water consumption, are borne by the surrounding community.

In regions where water is a critical resource for agriculture, increased competition from data centers can affect livelihoods. Farmers may face reduced water availability, leading to lower crop yields and economic losses. Similarly, households may experience higher water prices or restrictions on usage, affecting quality of life.

These dynamics can lead to social tensions and conflicts between data center operators and local communities. In some cases, public opposition has emerged in response to concerns about water use and environmental impact. This highlights the importance of considering social acceptance and community engagement in data center planning and operation.

5.5 Regional Disparities and Global Inequality

The environmental and socio-economic impacts of data centers are not evenly distributed across regions. Developed regions with advanced infrastructure and regulatory frameworks may be better equipped to manage the impacts of data center operations. In contrast, developing regions may face greater challenges due to limited resources and weaker governance structures.

Global demand for AI services is driven by users and organizations across the world, but the physical infrastructure supporting these services is concentrated in specific locations. This creates a spatial disconnect between the benefits of AI and its environmental costs. Regions hosting data centers may experience increased water stress and ecological impact, while the benefits of digital services are distributed globally.

This disparity raises important questions about equity and responsibility in the digital economy. It suggests the need for mechanisms to ensure that the costs and benefits of AI infrastructure are more fairly distributed, both within and across countries.

5.6 Governance, Policy, and Regulatory Challenges

The governance of data center water use presents significant challenges. Existing regulatory frameworks often focus on energy consumption and emissions, with less attention given to water usage. As a result, water-related impacts may not be adequately addressed in planning and operational decisions.

In many regions, water use by data centers is subject to general industrial regulations rather than specific guidelines tailored to digital infrastructure. This can lead to gaps in oversight and inconsistent enforcement. In addition, the lack of standardized reporting requirements makes it difficult to assess and compare water usage across facilities.

Policy responses are beginning to emerge, including requirements for water disclosure, efficiency standards, and incentives for sustainable practices. However, these measures are still in early stages and vary widely across jurisdictions. There is a need for more comprehensive and coordinated policy frameworks that address the full range of environmental impacts associated with data centers.

Effective governance also requires collaboration between stakeholders, including government agencies, industry operators, and local communities. Transparent communication and data sharing are essential to build trust and support informed decision-making.

5.7 Toward Sustainable and Equitable AI Infrastructure

Addressing the environmental and socio-economic impacts of AI data center water use requires a shift toward more sustainable and equitable infrastructure models. This includes the adoption of water-efficient technologies, the integration of environmental considerations into site selection, and the development of policies that balance economic growth with resource conservation.

In addition, there is a need to incorporate social considerations into sustainability strategies. This involves engaging with local communities, assessing potential impacts, and ensuring that benefits are shared more equitably. It also requires recognizing that sustainability is not only a technical challenge but also a social and political one.

The complexity of these issues underscores the importance of interdisciplinary approaches that combine technical innovation with policy development and social engagement. By adopting such approaches, it is possible to mitigate the negative impacts of AI infrastructure while maximizing its benefits.

5.8 Industry Case Examples and Regional Insights

Recent industry developments provide concrete evidence of the growing tension between AI expansion and water sustainability. Major hyperscale operators have reported significant water usage associated with AI workloads, particularly in regions with high cooling demand. For example, data centers located in arid regions have faced public scrutiny due to their reliance on evaporative cooling, which competes with local agricultural and residential water needs.

In developing regions, the challenges are even more pronounced due to limited infrastructure and weaker regulatory frameworks. Data centers are often attracted by low energy costs and favorable policies, but water availability is not always considered in site selection. This creates a mismatch between economic incentives and environmental capacity, increasing the risk of long-term sustainability issues.

These real-world examples highlight the need for context-aware planning and policy intervention. They also demonstrate that sustainability challenges are not purely technical but are shaped by regional conditions, governance structures, and socio-economic dynamics.

The next section builds on this analysis by proposing practical methodologies and design strategies to reduce water consumption and improve the sustainability of AI data centers.

6. DESIGN, IMPLEMENTATION, AND OPTIMIZATION OF SUSTAINABLE AI DATA CENTER COOLING SYSTEMS

6.1 Engineering Design Specifications for AI Data Centers

The design of sustainable AI data centers requires a shift from conventional facility planning toward performance-driven engineering that integrates computational intensity, thermal dynamics, and water-resource constraints. Unlike traditional enterprise data centers, AI-oriented facilities are characterized by high-density workloads, often exceeding 30–80 kW per rack and, in extreme cases, surpassing 100 kW per rack for advanced GPU clusters. This escalation in power density fundamentally transforms cooling requirements and introduces significant implications for both energy and water consumption. Figure 4 shows the Integrated framework for sustainable AI data center design.

Figure 4. Integrated framework for sustainable AI data center design.

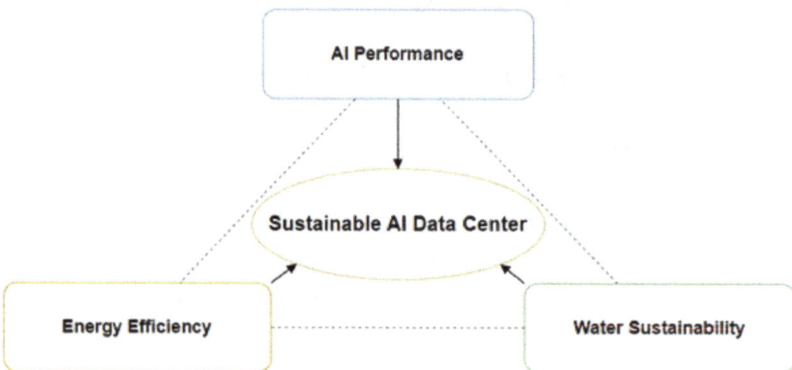

A critical design parameter is the relationship between power density and heat rejection. Approximately all electrical energy consumed by computing hardware is converted into heat, which must be removed continuously to maintain operational stability. As a result, thermal design becomes the central constraint in infrastructure planning. Engineers must define cooling capacity not only in terms of peak load but also in terms of dynamic variability associated with AI workloads, which can fluctuate rapidly depending on training cycles and inference demand.

Water usage effectiveness (WUE) has emerged as a key metric alongside power usage effectiveness (PUE). While PUE measures energy efficiency, WUE captures the volumetric water consumption per unit of IT energy. Sustainable design requires simultaneous optimization of both metrics, as improvements in energy efficiency may inadvertently increase water consumption in evaporative systems. Therefore, design specifications must incorporate multi-objective optimization criteria that balance energy, water, and computational performance.

Site selection also plays a decisive role in engineering design. Regions with cooler climates enable the use of economization strategies, reducing reliance on water-intensive cooling. Conversely, arid or water-stressed regions necessitate alternative approaches such as closed-loop liquid cooling or dry cooling systems. The integration of climate data, water availability indices, and regulatory constraints into early-stage design decisions is essential for minimizing long-term environmental impact.

6.2 Cooling System Selection Framework

Cooling technology selection requires a structured decision framework that considers workload characteristics, environmental conditions, and sustainability objectives, as no single solution is universally optimal. Each cooling approach involves trade-offs in energy efficiency, water consumption, cost, and operational complexity. Air cooling remains widely used due to its simplicity and lower cost, but its effectiveness declines at high power densities. Evaporative cooling improves energy efficiency but increases water consumption, which can create sustainability challenges in water-scarce regions.

More advanced solutions, such as direct-to-chip liquid cooling and immersion cooling, provide higher thermal efficiency and better suitability for high-density AI workloads. Liquid cooling reduces thermal resistance and can operate in closed-loop systems to minimize water loss, while immersion cooling offers superior performance with minimal water dependency but faces challenges in scalability and maintenance. Therefore, selecting an appropriate cooling system requires a multi-criteria evaluation that considers thermal performance, water use, climate suitability, scalability, and lifecycle cost, supported by decision-making tools to optimize deployment.

Selecting an appropriate cooling system is not only a technical decision but also a strategic one that affects operational cost, environmental impact, and long-term scalability. Decision-makers must evaluate trade-offs between energy efficiency and water consumption, particularly in regions with water scarcity or regulatory constraints.

From a managerial perspective, hybrid cooling strategies offer practical advantages by allowing flexibility across varying workloads and environmental conditions. Organizations should also consider lifecycle cost analysis, risk assessment, and regulatory compliance when selecting cooling technologies, rather than relying solely on initial capital expenditure or energy efficiency metrics.

6.3 Practical Implementation Methodology

The transition from conceptual design to operational deployment requires a structured implementation methodology that integrates engineering analysis, environmental assessment, and continuous optimization to ensure sustainability objectives are achieved. This process begins with workload characterization to estimate power consumption and thermal output, followed by thermal modeling and simulation using tools such as computational fluid dynamics and digital twins to optimize airflow, heat distribution, and system performance before deployment.

Subsequent steps include cooling system integration with appropriate infrastructure, water resource assessment to evaluate availability and environmental risks, and deployment with commissioning to validate operational performance. Finally, continuous monitoring and optimization through sensors, data analytics, and AI-driven control systems enable dynamic adjustments that improve efficiency, reduce energy and water consumption, and maintain long-term system reliability.

6.4 AI-Driven Optimization and Adaptive Cooling

Artificial intelligence enhances the efficiency of data center cooling systems by enabling adaptive control based on real-time data and predictive analytics. Sensor data on temperature, humidity, airflow, and equipment utilization are processed using machine learning models to predict thermal behavior, allowing dynamic adjustment of cooling parameters such as fan speed, coolant flow, and temperature setpoints. This approach improves system responsiveness compared to traditional static control methods.

AI-driven systems also enable multi-objective optimization, balancing energy efficiency, water conservation, and operational reliability. In addition, predictive maintenance capabilities allow early detection of anomalies and potential equipment failures, reducing downtime and extending infrastructure lifespan. AI can further support demand-response strategies by aligning cooling operations with energy availability, particularly in renewable-powered facilities, thereby improving both sustainability and operational flexibility.

6.5 Integrated Design for Ecological Balance

Achieving ecological sustainability in AI data centers requires an integrated design approach that aligns technological innovation with environmental and social considerations, where cooling systems are treated as part of a broader resource ecosystem encompassing energy, water, and community impacts. Hybrid cooling architectures offer a balanced solution by combining technologies, such as using air cooling under moderate conditions and liquid cooling during peak loads, while water reuse systems reduce freshwater demand through recycling and treatment. In this context, circular resource management becomes essential, enabling the reuse of waste heat for district heating or industrial applications and the recovery and reuse of cooling water to minimize environmental impact. At the same time, policy and governance frameworks play a crucial role by shaping design and operational practices through regulations on water use, emissions, and reporting, with effective sustainability requiring collaboration among industry, government, and local communities. Table 2 shows the Synthesis of cooling and sustainability strategies across scenarios and associated risks.

Table 2. Synthesis of cooling and sustainability strategies across scenarios and associated risks.

Method	Scenario	Benefit	Risk
Evaporative Cooling	High heat, low humidity	Energy efficient	High water usage
Liquid Cooling	High-density AI racks	Thermal efficiency	High cost and complexity
Immersion Cooling	Extreme compute clusters	Minimal water use	Infrastructure change
Hybrid Cooling	Variable workloads	Flexibility	System integration challenges
Water Recycling	Water-scarce regions	Reduced freshwater use	Treatment cost
AI Optimization	Dynamic workloads	Efficiency gains	Model dependency
Climate-Aware Siting	Planning phase	Long-term sustainability	Limited location options

Ultimately, the integration of engineering design, advanced cooling technologies, AI-driven optimization, and environmental stewardship defines the future of sustainable AI data centers. By adopting a holistic approach, it is possible to mitigate the ecological footprint of AI infrastructure while supporting continued technological advancement.

7. ANALYSIS AND DISCUSSION

The analysis highlights that sustainability challenges in AI data centers are not isolated technical problems but systemic issues that arise from the interaction between computation, energy, and water resources. Understanding these interactions is essential for developing effective strategies that balance performance with environmental responsibility. For decision-makers, these findings imply that sustainability strategies must move beyond isolated efficiency improvements toward integrated resource management.

7.1 Interdependence of Computation, Energy, and Water

A key finding is the strong interdependence between computational demand, energy consumption, and water usage in AI data centers, forming a tightly coupled system where changes in one dimension directly affect the others. The growth of AI workloads increases computational demand, which drives higher energy consumption and consequently generates more heat that must be dissipated through cooling systems. In many cases, especially with evaporative cooling, this thermal demand is translated into increased water consumption, creating a cascading resource dependency.

This interconnected relationship challenges traditional optimization approaches that focus on a single resource. Improvements in energy efficiency may lead to increased water use, while efforts to reduce water consumption, such as dry cooling, can raise energy demand and emissions. Therefore, sustainability in AI data centers must be addressed through integrated, multi-dimensional strategies that balance energy, water, and computational performance.

7.2 Trade-Offs Across Cooling Technologies

The analysis of cooling technologies highlights inherent trade-offs that influence both design and operational decisions. Air-based systems offer simplicity and compatibility but struggle to support high-density workloads, while evaporative cooling improves energy efficiency at the cost of higher water consumption. More advanced solutions such as liquid and immersion cooling provide superior thermal performance and potential water savings but involve higher capital costs and greater operational complexity.

These trade-offs are context-dependent and vary based on climate, workload characteristics, and infrastructure conditions. For instance, evaporative cooling may be suitable in water-abundant regions, whereas liquid cooling may be preferable in water-scarce environments. As a result, decision-making must move beyond single-objective optimization and adopt holistic frameworks that consider environmental, economic, and operational factors simultaneously.

7.3 Regional Context and Uneven Impact Distribution

The environmental impact of AI data centers on water resources varies significantly across regions, influenced by climate conditions, hydrological characteristics, and governance frameworks. In water-rich regions, the effects of data center water consumption may be relatively manageable, particularly when supported by sustainable management practices. However, in water-scarce areas, even moderate water use can place significant pressure on local communities and ecosystems.

This uneven distribution of impacts creates a spatial imbalance between the global benefits of AI and the localized environmental costs. While users worldwide benefit from AI services, the resource burdens are concentrated in specific regions. Addressing this imbalance requires policy interventions and improved planning strategies, including the integration of regional water availability into site selection and infrastructure development decisions.

7.4 Limitations of Current Metrics and Reporting Practices

The assessment of data center sustainability is limited by existing metrics and reporting practices, which do not fully capture the complexity of environmental impacts. Metrics such as Power Usage Effectiveness and Water Usage Effectiveness provide useful benchmarks but remain narrow in scope. Power Usage Effectiveness focuses on internal energy efficiency without accounting for upstream emissions or water use, while Water Usage Effectiveness does not consider regional water scarcity or ecological consequences.

In addition, the lack of standardized reporting frameworks makes it difficult to compare performance across data centers. Variations in methodology, system boundaries, and disclosure practices reduce transparency and hinder accurate evaluation. This limitation restricts the ability of researchers and policymakers to assess the true environmental footprint of AI infrastructure, highlighting the need for more comprehensive and standardized sustainability metrics. Table 3 shows the Comparison of sustainability metrics for data center evaluation.

Table 3. Comparison of sustainability metrics for data center evaluation.

Metric	Focus	Strength	Limitation
PUE	Energy efficiency	Simple benchmark	Ignores water impact
WUE	Water efficiency	Captures water use	No regional context
CUE	Carbon emissions	Climate relevance	Ignores water
Integrated Index (Proposed)	Energy + Water + Carbon	Holistic	Requires standardization

These limitations highlight the need for more comprehensive and context-aware metrics. Future frameworks should integrate multiple dimensions of sustainability, including energy, water, carbon emissions, and social impact. Such metrics would enable more informed decision-making and facilitate the development of effective policies.

7.5 Role of Innovation and Digital Optimization

While technological innovation offers significant potential, its effectiveness depends on practical implementation and organizational readiness. AI-driven optimization systems can reduce resource consumption, but their deployment requires investment in data infrastructure, skilled personnel, and operational integration.

In practice, organizations must combine technological innovation with governance mechanisms and behavioral changes. This includes adopting water-efficient practices, improving transparency, and aligning sustainability goals with business strategies. Without such integration, technological solutions alone are unlikely to deliver meaningful long-term impact.

7.6 Governance and Policy Implications

Governance and policy play a critical role in shaping the sustainability of AI data centers, as market-driven approaches may prioritize short-term economic gains over long-term environmental considerations. Policy interventions such as water usage standards, mandatory reporting, and incentives for sustainable technologies can guide more responsible practices. In addition, land-use planning and zoning regulations can help ensure that data centers are located in regions with adequate resource capacity.

Given the global nature of AI infrastructure, international cooperation is also necessary to address the uneven distribution of environmental impacts. Effective governance requires collaboration among governments, industry, and local communities, supported by transparency and stakeholder engagement. Such coordinated efforts are essential to balance technological development with environmental sustainability and social equity.

7.7 Synthesis and Contribution

An integrated analysis of water consumption in AI data centers is presented by linking technical, environmental, and socio-economic perspectives, addressing the limitations of existing approaches that focus primarily on energy efficiency and highlighting water as a critical yet underexplored dimension of sustainability. The interactions between cooling technologies, environmental impacts, and governance frameworks are examined to provide a more comprehensive understanding of AI infrastructure sustainability. Table 4 shows the Key Trade-Offs in AI Data Center Sustainability.

Table 4. Key Trade-Offs in AI Data Center Sustainability

Dimension	Benefit	Risk	Implication
Energy Efficiency	Lower emissions	Higher water use	Trade-off management
Water Efficiency	Reduced water stress	Higher energy demand	Context-based choice
Advanced Cooling	High performance	High cost	Selective deployment
AI Optimization	Efficiency gains	System dependency	Requires governance

Practical insights are also offered through proposed design strategies and implementation methodologies aimed at reducing water consumption and improving overall system efficiency. These contributions respond directly to gaps in the literature and emphasize the importance of applied, interdisciplinary approaches that support both technological advancement and sustainable development.

8. CONCLUSION

The rapid expansion of artificial intelligence has transformed data centers into critical digital infrastructure, but this growth has also increased environmental pressure, particularly in relation to water consumption. This chapter shows that sustainability in AI data centers cannot be evaluated only through energy efficiency or carbon emissions, since cooling systems create a strong interdependence between computation, energy, and water usage. As AI workloads continue to grow, cooling demand and associated water consumption are likely to increase significantly, making water a central sustainability concern rather than a secondary operational issue.

The analysis also highlights important trade-offs across cooling technologies. Air-based systems remain simple and cost-effective but struggle with high-density AI workloads, while evaporative cooling improves energy efficiency at the expense of high water use. More advanced technologies, such as liquid cooling and immersion cooling, provide better thermal performance and lower water dependency, although they require higher investment and operational complexity. These findings demonstrate that no single cooling technology is universally optimal, and future AI data centers will require flexible, hybrid, and context-aware cooling strategies adapted to environmental and operational conditions.

From an environmental and socio-economic perspective, the growth of AI infrastructure creates broader challenges related to water stress, ecological impact, and unequal resource distribution. In many regions, data center operations compete with local communities and ecosystems for limited water resources, especially in areas already facing water scarcity. Existing sustainability metrics and reporting practices remain limited because they often fail to capture regional water conditions and the combined impacts of energy, water, and carbon use. This indicates the need for more integrated frameworks that support transparency, accountability, and long-term environmental planning.

To address these challenges, the chapter proposes integrated design and optimization approaches that combine advanced cooling technologies, AI-driven thermal management, water recycling, climate-aware site selection, and stronger governance frameworks. Sustainable AI infrastructure will depend not only on technological innovation but also on policy development, interdisciplinary collaboration, and responsible resource management. By integrating technical, environmental, and socio-economic perspectives, this chapter contributes to a more comprehensive understanding of how AI data centers can evolve toward sustainable and equitable digital infrastructure in an increasingly resource-constrained world.

REFERENCES

Al Kez, D., Foley, A. M., Laverty, D., Del Rio, D. F., & Sovacool, B. (2022). Exploring the sustainability challenges facing digitalization and internet data centers. *Journal of Cleaner Production*, *371*, 133633. Advance online publication. DOI: 10.1016/j.jclepro.2022.133633

Brodie, P. (2023). Data infrastructure studies on an unequal planet. *Big Data & Society*, *10*(1), 20539517231182402. Advance online publication. DOI: 10.1177/20539517231182402

Cao, K., Li, Z., Luo, H., Jiang, Y., Liu, H., Xu, L., Gao, P., & Liu, H. (2024). Comprehensive review and future prospects of multi-level fan control strategies in data centers for joint optimization of thermal management systems. *Journal of Building Engineering*, *94*, 110021. Advance online publication. DOI: 10.1016/j.jobe.2024.110021

Chen, C., Ye, C., Ding, Y., Xiong, K., & Tang, X. (2026). Exploring the Demand Response Potential of Liquid-Cooled Data Centers. *Applied Energy*, *412*, 127679. Advance online publication. DOI: 10.1016/j.apenergy.2026.127679

Corigliano, O., Algieri, A., & Fragiacomo, P. (2024). Turning Data Center Waste Heat into Energy: A Guide to Organic Rankine Cycle System Design and Performance Evaluation. *Applied Sciences (Basel, Switzerland)*, *14*(14), 6046. Advance online publication. DOI: 10.3390/app14146046

d'Orgeval, A., Sheehan, S., Avenas, Q., Assoumou, E., & Sessa, V. (2026). Generative AI impact assessment through a life cycle analysis of multiple data center typologies. *Applied Energy*, *406*, 127288. Advance online publication. DOI: 10.1016/j.apenergy.2025.127288

Fu, Z.-L., Cao, C., & Gao, F. (2025). Impact of artificial intelligence energy management technologies on commercial multi-energy consumption. *Energy*, *340*, 139247. Advance online publication. DOI: 10.1016/j.energy.2025.139247

Jiang, H., Li, N., Zhao, C., & Shen, L. (2026). Research on return water temperature prediction model for casting cooling system based on deep learning. *Scientific Reports*, *16*(1), 4432. Advance online publication. DOI: 10.1038/s41598-025-34510-5 PMID: 41501340

Lambert, S., Caron, E., Lefevre, L., & Grivel, R. (2026). Consolidation of virtual machines to reduce energy consumption of data centers by using ballooning, sharing and swapping mechanisms. *Future Generation Computer Systems*, *174*, 107968. Advance online publication. DOI: 10.1016/j.future.2025.107968

Lin, J., Lin, W., Huang, H., Lin, W., & Li, K. (2024). Thermal Modeling and Thermal-Aware Energy Saving Methods for Cloud Data Centers: A Review. *IEEE Transactions on Sustainable Computing*, *9*(3), 571–590. DOI: 10.1109/TSUSC.2023.3346332

Long, N. V., Vinh, T. N., Quyen, H. V., Duyen, N. T. M., Hoa, P. V., & Tuan, M. A. (2026). The Evolution of AI Chips: From Cloud to Edge, Powering the Future of Artificial Intelligence. In *Lecture Notes of the Institute for Computer Sciences, Social-Informatics and Telecommunications Engineering, LNICST* (Vol. 673 LNICST, pp. 188–217). DOI: 10.1007/978-3-032-14055-5_14

Ma, S., Liu, Y., Liu, Y., Wang, J., Fang, Q., & Huang, Y. (2025). Artificial intelligence-enabled predictive energy saving planning of liquid cooling system for data centers. *Advanced Engineering Informatics*, *65*, 103283. Advance online publication. DOI: 10.1016/j.aei.2025.103283

Rianisa, A., Anggraini, D. S., & Sekti, B. A. (2026). Role of AI in Consumer Decision-Making Processes. In *Modern Consumer Behavior at the Intersection of AI and Social Media* (pp. 123–151). DOI: 10.4018/979-8-3373-2747-1.ch005

Sekti, B. A. (2025). AI Technology Management Concepts and Strategies. In *AI Technology Management and Value Creation* (pp. 1–32). DOI: 10.4018/979-8-3373-2497-5.ch001

Sekti, B. A. (2026). Cybersecurity Challenges in Smart City Development. In *Securing Smart Cities Through Modern Cryptography Technologies* (pp. 75–102). IGI Global Scientific Publishing., DOI: 10.4018/979-8-3373-3326-7.ch003

Shahi, P., Heydari, A., Modi, H., Chinthaparthy, L. S. R., Barigala, A., Eslami, B., Tradat, M., Chowdhury, U., Agonafer, D., & Rodriguez, J. (2025). Liquid to Liquid In-Row Coolant Distribution Unit Characterization for High Power Data Center. *Journal of Electronic Packaging*, *147*(4), 041109. Advance online publication. DOI: 10.1115/1.4069348

Shi, W., Ma, X., Min, Y., & Yang, H. (2024). Feasibility Analysis of Indirect Evaporative Cooling System Assisted by Liquid Desiccant for Data Centers in Hot-Humid Regions. *Sustainability (Basel)*, *16*(5), 2011. Advance online publication. DOI: 10.3390/su16052011

Utami, A. S., & Sekti, B. A. (2025). Al in Critical Sectors: Healthcare, Education, Finance, and Security - Ethical Challenges in Generative Al. In *Navigating Social Trust in the AI Era* (pp. 155–186). DOI: 10.4018/979-8-3373-7011-8.ch007

Wan, J., Duan, Y., Gui, X., Liu, C., Li, L., & Ma, Z. (2023). SafeCool: Safe and Energy-Efficient Cooling Management in Data Centers with Model-Based Reinforcement Learning. *IEEE Transactions on Emerging Topics in Computational Intelligence*, *7*(6), 1621–1635. DOI: 10.1109/TETCI.2023.3234545

KEY TERMS AND DEFINITIONS

AI Data Center: A data center designed to support artificial intelligence workloads that require high computational power, advanced processors, and intensive cooling systems.

Cooling Technology: A system or method used to remove heat from servers and computing infrastructure in order to maintain stable operational temperatures.

Evaporative Cooling: A cooling approach that lowers temperature through water evaporation, commonly used in large-scale data center operations.

Green Data Center: A data center that applies environmentally sustainable practices such as energy efficiency, water conservation, and low-carbon operations.

Liquid Cooling: A cooling method that uses liquid instead of air to transfer heat away from high-performance computing equipment.

Resource Efficiency: The ability to minimize the use of energy, water, and materials while maintaining effective operational performance.

Sustainable Computing: The practice of designing and operating computing systems in ways that reduce environmental impact and improve resource efficiency.

Thermal Management: The process of controlling and optimizing heat generation and heat removal within computing infrastructure.

Water Usage Effectiveness: A sustainability metric used to measure the amount of water consumed by a data center relative to its energy usage.

Chapter 4
The Water Footprint of Artificial Intelligence:
Data Center Expansion and Freshwater Sustainability

Pratibha Agrawal
http://orcid.org/0009-0003-8820-9553
Amity University, Ranchi, India

Akanksha Yadav
http://orcid.org/0000-0002-8096-0161
Amity University, Ranchi, India

ABSTRACT

Artificial intelligence (AI) technology is rapidly growing, which has triggered a dramatic increase in the size of data centers around the world, attracting serious environmental issues. The chapter discusses the outcomes of this growth on the sustainability of freshwater, particularly the data center water-intensive operations. It relies on the available literature, on the basis of the available empirical data and assesses the trends in the data center proliferation, regional differences in the water consumption, and on their direct and indirect effects in regard to the freshwater resources. The paper also examines the impact of climatic, economic, and policy dynamics on water consumption, which increases pressure in susceptible areas. It suggests sustainable cooling technologies, governance systems, and corporate water stewardship as those areas of mitigation and presents the gaps in the existing literature and points to the necessity of integrated water resource governance in accordance with digital development.

DOI: 10.4018/979-8-2600-2761-5.ch004

INTRODUCTION

Digital technologies have become a defining feature of the Anthropocene, transforming economic production, governance, and everyday life through artificial intelligence (AI), cloud computing, and large-scale digital infrastructures. Although often perceived as immaterial, these technologies depend on extensive physical infrastructures, including data centers, energy systems, and cooling technologies that require substantial quantities of electricity, minerals, and freshwater. Recent scholarship has demonstrated that digital technologies are embedded within complex socio-ecological systems and generate significant environmental footprints (Masanet et al., 2020; Mytton, 2020, 2021). Studies further indicate that AI systems contribute substantially to freshwater demand through cooling requirements associated with model training and inference processes, as well as indirect water consumption linked to electricity generation (Li et al., 2024).

Artificial intelligence has emerged as one of the most transformative technological developments of the twenty-first century, reshaping economic systems, governance, healthcare, scientific research, and everyday social interactions. The rapid expansion of machine learning, deep learning, and generative AI applications has significantly increased demand for computational power and digital infrastructure. Consequently, hyperscale data centers have become essential components of the global digital economy, enabling the storage, processing, and transmission of vast quantities of data (Masanet et al., 2020). The rapid growth of cloud computing and digital services has further increased the scale and resource demands of data centers (Jones, 2018; Meijer, 2010).

While the environmental implications of AI have largely been examined through the lens of energy consumption and carbon emissions (Strubell et al., 2019), its water footprint has received comparatively less scholarly attention. Freshwater is a finite resource that is essential for ecosystems, agriculture, industry, and human well-being. Data centers depend on cooling technologies such as evaporative cooling and chilled-water systems to regulate the temperatures generated by high-performance computing equipment. These cooling processes can consume substantial quantities of freshwater, particularly in regions already experiencing hydrological stress (Mytton, 2020, 2021). Recent research demonstrates that AI model development and deployment require significant water resources through both direct cooling processes and indirect energy production systems (Ren et al., 2023; Li et al., 2025). Moreover, the environmental footprint of AI extends beyond operational cooling systems to include lifecycle water consumption associated with semiconductor manufacturing, hardware production, and resource extraction (Williams et al., 2002; Nagapurkar et al., 2023). Together, these interconnected processes reveal a broader water–energy–digital infrastructure nexus shaping contemporary technological systems.

This chapter argues that the rapid expansion of AI infrastructure, particularly hyperscale data centers, is creating new sustainability challenges through increasing pressure on freshwater resources. Although AI is widely promoted as a driver of innovation and digital transformation, its supporting infrastructures consume significant quantities of water through cooling systems, electricity generation, and hardware production processes. Furthermore, these environmental impacts are unevenly distributed geographically. Data centers are increasingly being established in regions already experiencing water scarcity, raising important questions concerning environmental justice, resource governance, and the unequal ecological costs of digital transformation.

The chapter addresses the following research question: What is the impact of the rapid expansion of artificial intelligence infrastructure, particularly hyperscale data centers, on freshwater consumption and sustainability at the global scale? To answer this question, the chapter examines the mechanisms through which AI systems generate water demand, explores regional variations in data center water use, and evaluates governance frameworks, technological innovations, and policy interventions aimed at reducing the water footprint of digital infrastructure. By synthesizing literature from environmental studies, digital infrastructure research, and sustainability governance, the chapter conceptualizes the water footprint of AI as a multidimensional phenomenon encompassing direct cooling requirements, indirect energy-related water use, and lifecycle impacts associated with hardware production and disposal. In doing so, it contributes to growing debates on sustainable digitalization and highlights the importance of integrating freshwater sustainability into future AI governance and infrastructure planning.

2. LITERATURE REVIEW AND THEORETICAL FRAMEWORK

The rapid growth of artificial intelligence (AI) has increased demand for high-performance computing infrastructure and hyperscale data centers. As Masanet et al. (2020) note, data centers have become critical components of the global digital ecosystem, supporting cloud computing, online services, and AI-driven applications. While digital infrastructure contributes to technological innovation and economic development, it also generates significant environmental impacts through resource consumption and infrastructure expansion.

Research on the environmental implications of AI has traditionally focused on energy consumption and carbon emissions associated with large-scale computing systems. Training advanced machine learning models requires substantial computational resources and electricity (Strubell et al., 2019). Data centers have consequently emerged as important contributors to global energy demand (Masanet et al., 2020).

However, recent scholarship emphasizes that the environmental footprint of AI extends beyond energy use to include freshwater consumption, electronic waste, mineral extraction, and supply-chain impacts (Chouksey et al., 2026). AI-intensive workloads increase cooling requirements and associated water use (Lei et al., 2025), while semiconductor manufacturing depends on large quantities of ultrapure water (Williams et al., 2002).

Increasing attention is therefore being paid to the water footprint of digital infrastructure. Data centers generate substantial heat and rely on cooling technologies, particularly evaporative cooling systems, that consume significant quantities of freshwater (Mytton, 2020, 2021; de Vries-Gao, 2025). Mytton (2021) identifies Water Usage Effectiveness (WUE) as an important metric for evaluating the sustainability of data center operations. The concept of the water footprint encompasses both direct and indirect water consumption embedded within production systems and supply chains (Hoekstra et al., 2011a, 2011b). Studies demonstrate that AI model training and deployment involve measurable water use through cooling requirements and electricity generation processes (Ren et al., 2023; Li et al., 2023). As AI adoption expands, the growing water footprint of digital infrastructure presents important sustainability challenges, particularly in water-stressed regions.

AI is frequently promoted as a tool for addressing sustainability challenges through applications in climate monitoring, energy optimization, agricultural efficiency, and water resource management (Vinuesa et al., 2020). Nevertheless, scholars argue that these benefits must be evaluated alongside environmental costs. Chouksey et al. (2025) describe a sustainability paradox in which AI simultaneously creates environmental opportunities and ecological burdens. Consequently, improving algorithmic efficiency, optimizing data center operations, and integrating renewable energy sources have become important sustainability strategies (IEA, 2023). Environmental governance has also gained prominence, with governments and technology firms developing initiatives to improve energy efficiency, reduce emissions, and enhance water stewardship. However, limited transparency regarding water consumption remains a significant challenge. Privette, Barros, and Cai (2026) emphasize that inadequate disclosure of data center water use constrains effective regulation and sustainability planning, while Mytton (2021) notes that limited access to detailed water-use data hampers environmental assessment.

This chapter adopts an integrated theoretical framework combining the water–energy nexus, sustainable digitalization, and environmental justice perspectives. The water–energy nexus highlights the interdependence of water and energy systems (Hoff, 2011), while Hlabisa (2025) conceptualizes AI as a socio-metabolic infrastructure embedded within broader ecological systems. Sustainable digitalization emphasizes aligning technological innovation with environmental sustainability (Vinuesa et al., 2020), whereas environmental justice perspectives draw attention to the unequal distribution of environmental burdens associated with data center expansion. Methodologically, the chapter employs a qualitative research design based on a systematic review of interdisciplinary literature and secondary sources obtained from Scopus, Web of Science, ScienceDirect, and Google Scholar. The literature was analyzed through thematic categorization focusing on water-consumption pathways, regional sustainability challenges, and governance mechanisms associated with AI infrastructure and freshwater sustainability.

3. CONCEPTUAL FRAMEWORK LINKING AI EXPANSION AND FRESHWATER SUSTAINIBILITY

The rapid growth of artificial intelligence (AI) technologies has accelerated the expansion of digital infrastructure and increased demand for high-performance computing systems. These systems are primarily housed in large-scale data centers responsible for processing, storing, and analyzing vast volumes of digital information. While AI technologies offer significant opportunities for technological innovation and economic development, they also generate important environmental challenges, particularly concerning freshwater consumption. Understanding the relationship between AI expansion and water sustainability is therefore essential for assessing the environmental implications of digital transformation.

The conceptual framework developed in this chapter highlights the interconnections among AI adoption, data center expansion, cooling requirements, freshwater consumption, and sustainability governance. As AI applications become increasingly integrated into sectors such as healthcare, finance, manufacturing, and transportation, demand for computational capacity continues to rise. This growth has stimulated the development of hyperscale data centers capable of supporting advanced machine learning models, cloud computing services, and large-scale data analytics.

Data centers operate continuously and generate substantial amounts of heat, necessitating sophisticated cooling systems that often depend on freshwater resources. Consequently, the expansion of AI infrastructure contributes to water consumption through multiple pathways, including direct water use for cooling operations, indirect water use associated with electricity generation, and lifecycle water consumption linked to hardware manufacturing and disposal. These interconnected processes illustrate the complex relationship between technological advancement and environmental sustainability. Building upon these theoretical insights, the conceptual framework illustrates the links among AI expansion, data centre growth, freshwater consumption, environmental impacts, and sustainability governance (see Figure 1).

Figure 1. Conceptual Framework Linking

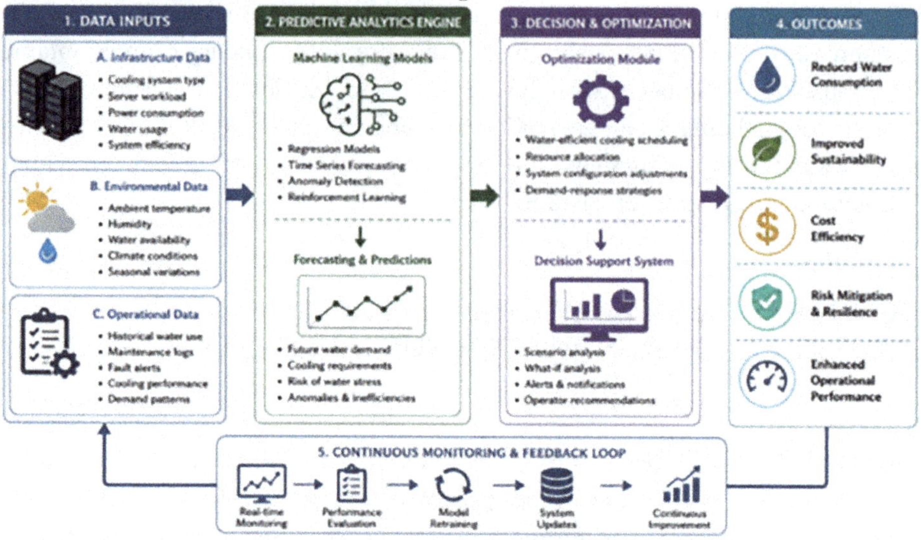

Note: The framework integrates multi-source data, predictive analytics, and optimization strategies to enhance water-use efficiency and sustainability in AI data center operations.

The rapid expansion of AI data centers is driven by the growing demand for cloud computing services and computationally intensive applications such as deep learning, predictive analytics, and natural language processing. These technologies require large-scale data storage and processing capabilities, prompting technology companies to invest heavily in hyperscale data centers. The digitalization of industries has further accelerated this trend, as businesses increasingly rely on AI-driven systems to improve operational efficiency, decision-making, and service delivery. In addition, government initiatives promoting digital economies and artificial intelligence innovation have encouraged investment in large-scale digital infrastructure. Consequently, data centers have become critical components of national technological and economic development strategies.

The conceptual framework identifies three principal pathways through which AI infrastructure affects freshwater resources. The first is direct water consumption associated with cooling systems. Data centers generate substantial heat during computational processes and therefore require continuous cooling, often through evaporative cooling and chilled-water systems that consume significant quantities of freshwater. The second pathway is indirect water consumption linked to electricity generation. AI data centers require large amounts of electricity, and in many regions power generation depends on water-intensive thermal power plants. The third pathway is lifecycle water consumption, which arises from the manufacturing and disposal of digital hardware. Semiconductor fabrication, electronic component production, and raw-material extraction require substantial volumes of water throughout their production processes. Together, these pathways demonstrate that the water footprint of AI extends beyond operational activities and encompasses the entire lifecycle of digital infrastructure.

Several factors can moderate the environmental impacts of AI data centers and improve water sustainability. Technological innovation plays a crucial role in reducing the water footprint of digital infrastructure. Advanced cooling technologies, including air-based cooling and liquid immersion cooling, are generally more water-efficient than conventional evaporative cooling systems. The adoption of renewable energy sources also contributes to water sustainability. Compared with traditional thermal power plants, solar and wind energy require considerably less water for electricity generation. Consequently, transitioning AI infrastructure to renewable energy can significantly reduce its indirect water footprint.

Policy frameworks and regulatory mechanisms further influence water-consumption patterns in data center operations. Governments can promote water-efficient infrastructure through environmental regulations, reporting requirements, and sustainability standards that encourage responsible resource management. In addition, corporate sustainability initiatives, including water-conservation programs, water-restoration projects, and environmental reporting practices, have become increasingly important in improving water stewardship within the technology sector. Together, these technological, regulatory, and corporate measures can help mitigate the freshwater impacts of expanding AI infrastructure.

4. MECHANISMS OF WATER CONSUMPTION IN ARTIFICIAL INTELLIGENCE INFRASTRUCTURE

The rapid growth of artificial intelligence (AI) technologies has significantly increased demand for large-scale digital infrastructure, particularly hyperscale data centers that support machine learning, cloud computing, and data analytics. Although AI is often perceived as a digital or immaterial technology, it relies on extensive physical infrastructures that consume substantial quantities of natural resources. Among these, freshwater plays a critical role in maintaining the operational stability and efficiency of data center systems. As computational demands continue to grow, understanding the mechanisms through which AI infrastructure consumes water has become increasingly important for evaluating the environmental sustainability of digital technologies.

One of the primary pathways of water consumption is through data center cooling systems. High-performance computing equipment, including servers, processors, and graphics processing units (GPUs), generates significant amounts of heat during AI training and deployment processes. To prevent system failures and maintain operational efficiency, data centers rely on advanced cooling technologies. Evaporative cooling systems remain among the most widely used approaches. These systems circulate water through cooling towers, where heat is dissipated through evaporation. While highly effective, evaporative cooling requires continuous freshwater inputs because evaporated water cannot be recovered. Consequently, large-scale facilities can consume substantial volumes of water during routine operations (Mytton, 2021). Lei et al. (2025) further identify workload intensity, climatic conditions, and cooling-system configurations as important factors influencing operational water consumption in hyperscale data centers.

Another commonly used technology is chilled-water cooling, in which refrigerated water is circulated through heat exchangers to absorb heat from computing equipment. Although these systems provide effective temperature regulation, they also require continuous water and energy inputs. Water consumption varies according to facility size, computational intensity, and local climatic conditions, with facilities located in warmer regions generally requiring greater cooling resources. Beyond direct cooling requirements, AI infrastructure contributes to indirect water consumption through electricity generation. Data centers require large quantities of electricity to power computing devices, networking systems, and cooling equipment. In many regions, electricity is produced by thermoelectric power plants that consume significant amounts of water for steam generation and cooling processes. Consequently, the growth of AI infrastructure indirectly increases freshwater withdrawals associated with energy production. This relationship reflects the water–energy nexus, which emphasizes the interdependence of water and energy systems in modern economies (Hoff, 2011).

A third mechanism involves the lifecycle of digital hardware. AI systems depend on processors, GPUs, and semiconductor chips whose manufacturing requires substantial quantities of ultrapure water for wafer cleaning and fabrication processes. Williams, Ayres, and Heller (2002) demonstrate that semiconductor production is both energy- and water-intensive. In addition, the extraction and processing of raw materials such as lithium, copper, and rare-earth minerals require considerable water inputs during mining and refining operations. End-of-life management also contributes to environmental impacts, as electronic waste recycling and material recovery processes often require additional water resources.

Collectively, these mechanisms demonstrate that the water footprint of AI infrastructure extends across operational, manufacturing, and end-of-life stages. Samanta et al. (2025) emphasize that methodological choices significantly influence water-footprint estimation and sustainability assessments. Similarly, Lambert and Luccioni (2026) argue that environmental evaluations frequently overlook water use, materials manufacturing, and embodied infrastructure impacts, thereby underestimating the broader ecological consequences of AI systems.

These mechanisms are key to assessing the sustainability of digital infrastructure and hence the need to understand them. The researchers and policymakers will have a chance to evaluate the environmental trade-offs of digital technological development by pinpointing the main channels by which artificial intelligence systems use water. This type of knowledge can also serve as the basis of formulating policies that will help to minimize the water footprint of artificial intelligence infrastructure with technological progress, the creation of more efficient energy markets, and data centres with a more sustainable design. The principal mechanisms through which artificial intelligence infrastructure consumes freshwater are summarized in Table 1.

Table 1. Mechanisms of Water Consumption in Artificial Intelligence Infrastructure

Mechanism	Description	Water Use Process	Environmental Implication
Data Center Cooling Systems	Cooling systems regulate temperatures of servers and computing equipment used in AI operations.	Evaporative cooling and chilled water systems require continuous freshwater supply to dissipate heat.	High freshwater consumption, particularly in large hyperscale data centers.
Electricity Generation	AI systems require large amounts of electricity for data processing and storage.	Thermoelectric power plants use water for steam generation and cooling processes.	Indirect water consumption through energy production.
Hardware Manufacturing	AI infrastructure relies on advanced semiconductor chips and processors.	Semiconductor fabrication requires ultrapure water for cleaning silicon wafers.	Lifecycle water footprint associated with digital hardware production.
Raw Material Extraction	Mining of minerals used in electronic components such as copper and rare earth elements.	Water used for mineral separation, dust suppression, and chemical processing.	Environmental pressure on water resources in mining regions.

Source: Author's compilation based on Hoff (2011); Williams et al. (2002); Mytton (2021).

To provide a systematic understanding of the multidimensional pathways through which artificial intelligence infrastructure contributes to freshwater demand, the principal mechanisms of water consumption associated with AI systems are illustrated through direct and indirect water-use processes (See Figure 2).

Figure 2. Mechanisms of Water Consumption in AI Infrastructure

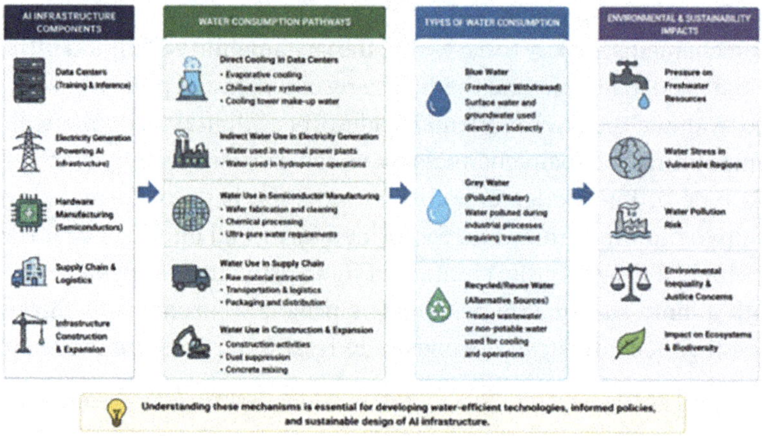

Source: Author's compilation based on Mytton (2021), Li et al. (2023), Williams et al. (2002), and Lei et al. (2025).

4.1 Global Expansion of Data Centers and Regional Water Stress

The rapid growth of artificial intelligence (AI) has accelerated the global expansion of hyperscale data centers that support cloud computing, data storage, and machine learning applications. While these facilities are essential to the digital economy, they have also increased concerns regarding freshwater consumption and environmental sustainability. Data centers require substantial amounts of electricity and cooling, making water use a critical issue, particularly in regions already experiencing water stress (Masanet et al., 2020).

The environmental impacts of data centers vary across geographical contexts. In the United States, large data center clusters in regions such as Virginia and Texas place increasing pressure on local water resources. Similar concerns have emerged in China, where many facilities are located in hydrologically stressed regions (Jiang et al., 2025). In arid regions of the Middle East, high cooling requirements further intensify freshwater demand. These patterns highlight important environmental justice concerns, as the benefits of AI are globally distributed while many environmental costs remain localized. Consequently, sustainable infrastructure planning, renewable energy integration, and water-efficient technologies are essential for reducing the environmental footprint of AI-driven digital infrastructure.

However, as the technologies of artificial intelligence are actively developing, it is possible that within the next several decades the demand in digital infrastructure will grow. The environmental impact of data center infrastructure will gain even more importance as AI applications are rolled out in sectors and societies alike. Knowing the geographic dynamics of digital infrastructure development and their engagement with the water systems of the region is thus critical to the establishment of the fact that the proliferation of artificial intelligence technologies is consistent with the goals of the long-term sustainability. Regional patterns of data center expansion and associated water-stress concerns are presented in Table 2.

Table 2. Regional Patterns of Data Center Expansion and Water Stress

Region	Major Data Center Locations	Key Drivers	Water Resource Concerns
North America	Virginia, California, Texas	High demand for cloud computing and AI infrastructure	High cooling demand and potential pressure on local water supplies
Europe	Ireland, Netherlands, Nordic countries	Access to renewable energy and strong digital infrastructure	Moderate water use; cooler climates reduce cooling demand
Asia	China, Singapore, India	Rapid digitalization and expansion of AI technologies	Water scarcity concerns in urban and industrial regions
Middle East	UAE, Saudi Arabia	Digital transformation and economic diversification	High cooling demand due to extreme climate conditions
Northern Europe	Sweden, Finland	Cooler climates and renewable energy resources	Lower water consumption due to free-air cooling systems

Source: Author's compilation based on Masanet et al. (2020); IEA (2023).

4.2 Governance, Sustainability Strategies, and Future Pathways for Water-Efficient AI Infrastructure

The rapid expansion of artificial intelligence (AI) infrastructure has increased concerns regarding its environmental impact, particularly freshwater consumption associated with data center operations. Addressing these challenges requires an integrated approach that combines technological innovation, corporate responsibility, and regulatory frameworks to promote sustainable digital infrastructure (Hiremath, 2024). Renewable energy integration, efficient cooling systems, and net-zero infrastructure strategies can further reduce the environmental footprint of AI servers (Xiao et al., 2025; Durmus et al., 2024; Manoharan et al., 2024).

A key strategy involves adopting water-efficient cooling technologies. Alternatives such as air-based cooling and liquid immersion cooling can significantly reduce dependence on water-intensive evaporative cooling systems while maintaining operational efficiency. The integration of renewable energy sources also contributes to sustainability by reducing both carbon emissions and the indirect water footprint associated with electricity generation. The International Energy Agency emphasizes the importance of energy efficiency and renewable energy adoption in reducing the environmental impacts of digital infrastructure (IEA, 2023). Similarly, Lei et al. (2025) demonstrate that combining renewable energy systems with water-efficient cooling technologies can substantially lower freshwater consumption.

Corporate sustainability initiatives, including water stewardship programs, water-positive commitments, and environmental reporting practices, have further enhanced accountability within the technology sector. However, voluntary measures alone may be insufficient. Governments play a crucial role in establishing environmental standards, water-efficiency regulations, and reporting requirements for data centers. Strategic infrastructure planning, particularly locating facilities in regions with cooler climates and greater renewable energy availability, can further reduce environmental pressures. Together, these approaches provide important pathways for ensuring that the continued growth of AI infrastructure aligns with long-term freshwater sustainability.

To illustrate the institutional and technological dimensions required to reduce the environmental footprint of artificial intelligence systems, a governance framework for sustainable AI infrastructure is proposed (see Figure 3).

Figure 3. Governance Framework for Sustainable AI Infrastructure

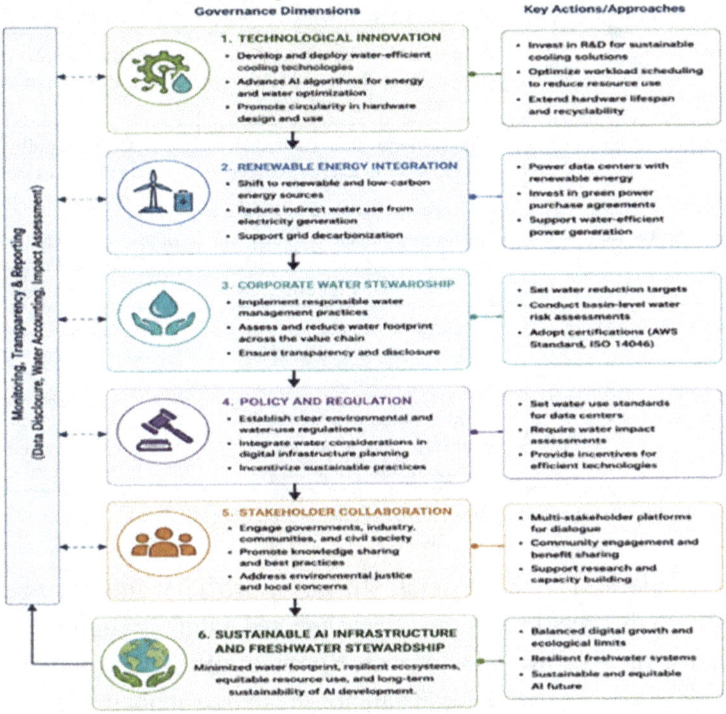

Source: Author's compilation based on OECD (2021), UN (2022), WBCSD (2023), and relevant literature.

The framework integrates technological innovation, renewable energy transition, corporate water stewardship, policy intervention, and stakeholder collaboration to support long-term freshwater sustainability.

The future directions of sustainable artificial intelligence infrastructure are probably to be a blend of technological empowerment, better governance structures as well as international collaboration . The environmental impact of digital infrastructure can be greatly minimized by investing in computing efficiency, cooling systems, and integration of renewable energy to minimize the environmental footprint. Simultaneously, it might be required that enhanced regulatory frameworks and global sustainability criteria are established so that the positive effects of artificial intelligence tools would not be related to the cost of environmental sustainability. Key sustainability strategies for reducing the water footprint of AI infrastructure are summarized in Table 3.

Table 3. Sustainability Strategies for Reducing the Water Footprint of AI Infrastructure

Strategy	Description	Expected Impact
Water-Efficient Cooling Technologies	Adoption of air cooling, liquid immersion cooling, and closed-loop cooling systems.	Reduces direct freshwater consumption in data centers.
Renewable Energy Integration	Transition to wind, solar, and other low-water energy sources.	Reduces indirect water consumption associated with electricity generation.
Corporate Water Stewardship	Technology companies implement water-positive initiatives and sustainability reporting.	Improves water conservation and environmental accountability.
Strategic Data Center Location	Locating data centers in cooler climates or water-abundant regions.	Reduces cooling demand and pressure on local water resources.
Policy and Regulatory Frameworks	Environmental regulations requiring water efficiency and sustainability reporting.	Encourages responsible management of digital infrastructure resources.

Source: Author's compilation based on IEA (2023); Vinuesa et al. (2020); Mytton (2021).

Ultimately, addressing the water footprint of artificial intelligence requires a holistic approach that recognizes the interconnected nature of digital infrastructure, energy systems, and freshwater resources. By integrating technological innovation with responsible governance and sustainable resource management strategies, it may be possible to align the continued growth of artificial intelligence technologies with broader environmental sustainability objectives.

4.3 Corporate Sustainability Initiatives in Artificial Intelligence Infrastructure

The rapid expansion of artificial intelligence (AI) infrastructure has increased concerns regarding the environmental impacts of hyperscale data centres, particularly their energy consumption, water use, and carbon emissions. In response, major technology firms have adopted sustainability initiatives aimed at reducing the ecological footprint of digital infrastructure. However, critics argue that efficiency improvements alone may not fully offset the environmental impacts associated with the continued expansion of AI systems and data centres.

Corporate sustainability strategies have become an important component of environmental governance within the technology sector. Many firms have introduced programs focused on renewable energy integration, water conservation, and resource efficiency within broader Environmental, Social, and Governance (ESG) frameworks. Among the most prominent initiatives are water-positive commitments. Google and Microsoft have pledged to achieve water-positive operations by 2030 through investments in water stewardship, restoration projects, and water-efficient cooling technologies.[1] Similarly, Amazon Web Services (AWS) has incorporated renewable energy investments, carbon-reduction strategies, and water-conservation measures into its sustainability agenda.

Technological innovation also plays a central role in these efforts. Alternative cooling technologies, including liquid immersion cooling and closed-loop water systems, can reduce dependence on water-intensive evaporative cooling. Many firms are additionally using recycled and reclaimed water to lessen pressure on freshwater resources. Sustainability reporting and environmental disclosure practices have further improved transparency and accountability regarding resource consumption.

Despite these advances, significant challenges remain. The increasing scale of AI infrastructure may continue to drive overall resource consumption, particularly in water-stressed regions. Consequently, stronger regulatory frameworks and collaborative governance approaches are needed to complement corporate sustainability initiatives and ensure the long-term environmental sustainability of AI infrastructure. Technology companies can invest to make the digital infrastructure development more sustainable by investing in water-efficient cooling technologies, renewable energy systems, and environmental reporting practices . Nevertheless, the long-term sustainability will not be achieved without cooperation between corporate entities, governments, and research institutions to make sure the intensive growth of artificial intelligence technologies does not put any extra strain on current environmental resources issues. Major corporate sustainability commitments adopted by leading technology companies are presented in Table 4.

Table 4. Sustainability Commitments of Major Technology Companies in Data Center Operations

Company	Key Sustainability Commitment	Water Management Strategy	Energy Strategy	Environmental Goal
Google	Water-positive operations by 2030	Use of recycled water, improved cooling technologies, watershed restoration projects	24/7 carbon-free energy across operations	Replenish more water than consumed in global operations
Microsoft	Water-positive commitment by 2030	Water replenishment projects, water-efficient cooling technologies	100% renewable energy by 2025	Reduce water consumption and support water restoration initiatives
Amazon (AWS)	Climate Pledge for net-zero carbon by 2040	Improved water efficiency in data centers and exploration of recycled water use	Investment in large-scale renewable energy projects	Achieve net-zero carbon emissions across operations
Meta (Facebook)	Sustainable data center infrastructure	Efficient cooling systems and water management strategies	100% renewable energy for global operations	Reduce environmental footprint of digital infrastructure

Source: Author's compilation based on corporate sustainability reports (Google Sustainability Report, Microsoft Environmental Sustainability Report, Amazon Climate Pledge, Meta Sustainability Report).

To sum up, corporate sustainability practices can be discussed as a significant part of the general initiative to minimize the environmental footprint of artificial intelligence infrastructure.

4.4 Environmental and Ethical Challenges of Artificial Intelligence Infrastructure

Despite the transformative potential of artificial intelligence (AI), the rapid expansion of AI infrastructure has generated significant environmental and ethical concerns. The growing reliance on hyperscale data centers, high-performance computing systems, and digital networks has increased pressure on energy and freshwater resources. Moreover, the environmental footprint of AI extends beyond operational activities to include impacts associated with hardware lifecycles, electronic waste generation, inference processes, and limited environmental transparency (Bozkurt, 2025).

A major challenge relates to the substantial computational resources required to train and operate advanced AI systems. Large language models and deep learning architectures depend on specialized hardware that generates considerable heat, necessitating extensive cooling systems that often rely on water-intensive technologies. Consequently, AI development contributes to increasing freshwater consumption, particularly where evaporative cooling systems are employed (Mytton, 2021). In addition, data centers require significant amounts of electricity, and in regions dependent on fossil-fuel-based power generation, expanding AI infrastructure may contribute to higher greenhouse gas emissions (Masanet et al., 2020).

AI infrastructure also raises concerns regarding environmental justice. While the benefits of AI technologies are globally distributed, the environmental burdens associated with maintaining data centers are often concentrated in specific localities, particularly water-stressed regions. Furthermore, limited transparency in environmental reporting remains a challenge, as detailed information on water consumption is often unavailable or inconsistent. Addressing these concerns requires stronger governance frameworks, improved transparency, sustainable cooling technologies, renewable energy integration, and greater corporate accountability. Ultimately, the long-term sustainability of AI depends on addressing the ecological and ethical implications of the infrastructures that support its operation (Bozkurt, 2025; Mytton, 2021; Masanet et al., 2020).

4.5 Global Case Studies of Data Center Water Consumption

The green aspects of artificial intelligence infrastructure are especially obvious when one considers regional trends in data center construction in various regions of the world. Since hyperscale data centers are getting larger to meet the increased computational needs of artificial intelligence systems, they will interact with local water systems in a wide variety of ways, which depend on regional climatic conditions, municipal infrastructure development, and water governance structures. The analysis of case studies in other geographic settings helps to gain in-depth information on the impact of the digital infrastructure growth on the sustainability of freshwater.

United States: Hyperscale Data Centers and Local Water Resources in

The United States is also home to one of the most concentrated data centers in the world and also a major infrastructure center of artificial intelligence. Some of the regions that have emerged to be key hubs in the development of digital infrastructure include Northern Virginia, California, Texas, and Arizona because of their high internet penetration, electricity accessibility, and association to key technology firms. The Northern Virginia region, commonly known as Data center alley, is commonly considered to be the largest data center cluster in the world with hundreds of data centers that are run by large technology companies and cloud service providers.

The high rate of expansion of data centers in this area has raised issues of water consumption in cooling systems. Data center sites with large sizes necessitate high amounts of water to maintain the temperatures in server rooms, especially when the computations are in full blast. The sustainability of the digital infrastructure expansion in the regions where water resources might be under growing pressure in response to climate variability and population growth has been questioned by local environmental organizations and policymakers.

To address such fears, some of the technology firms around the United States have started to embrace water-efficient cooling systems and alternative water sources. There are also facilities which recycle the wastewater or reclaimed water to cool facilities instead of using freshwater only. These efforts show that there has been an increasing awareness in the technology industry that sustainable water management is the key to the future sustainability of digital infrastructure.

China: Fast Digital growth and Environmental Stress

Another significant case study of artificial intelligence infrastructure expansion in the world is China. Within the last ten years, the Chinese government has financially invested heavily in the development of digital infrastructure as a larger initiative towards technological innovation and economic modernization. The rapidly developing digital economy in China has resulted in the development of many data centers used to facilitate cloud computing services, e-commerce, and artificial intelligence applications.

The high growth of the digital infrastructure, however, has also brought about environmental issues in China especially in the context of energy and water use. According to some studies, there are numerous Chinese data centers that are situated in the areas where the water shortage has already been observed. Most of the computing infrastructure in China is located in the north and western parts of the country, which are usually strapped by the arid climatic conditions with respect to availability of fresh water.

To overcome these difficulties, the Chinese government has unveiled policies that will enhance efficiency in the functioning of data centers and enhance more sustainable development of digital infrastructure. These policies involve prompting the migration of data centers to areas that have more access to renewable energy sources and those that have better climatic conditions. In other instances, data centers are built in areas with cooler temperatures in the north that will consume less water since the ambient temperatures will be low and will not require intensive cooling systems.

The experience of China shows that strategic planning of infrastructure is crucial in dealing with environmental effects of the development of artificial intelligence. Governments can contribute to the fact that the development of digital infrastructure will not unduly overload the natural resource system by taking into account the aspects of climate conditions, water availability, and energy resources, among other factors.

Europe: Sustainable Data Centers Design

The European nations have begun paying greater attention to building the digital infrastructure in a manner that would be more sustainable regarding environmental sustainability objectives. Countries like Sweden, Finland, Ireland and Netherlands are now becoming convenient places to build the data centres as they have a high level of digital connectivity, availability of renewable power, and comparatively cooler climates. Such aspects have enabled data centres in Europe to run more efficiently and minimise the use of cooling technologies that are water intensive.

Indeed, Nordic nationals like Sweden and Finland have encouraged the creation of energy efficient and environment friendly data centers by taking advantage of their cool climate and the vast amount of renewable energy they have. In such areas, there are usually the data-centers operating under the regime of free-air cooling, which involves the usage of the naturally cold air outside in order to cool the heat produced by the computing equipment. The method will help to massively decrease the use of water-consuming cooling systems and reduces the overall energy use.

Besides technological advancements, the European governments have engaged in regulatory measures that are aimed at enhancing sustainable development of digital infrastructure. The green deal and digital sustainability programs of the European Union address the fact of decreasing the environmental imprint of digital technologies, promoting the economic growth process and technological development. These policies motivate operators of data centers to invest in energy-saving technologies, increase environmental clarity and integrate renewable sources of energy into digital infrastructure infrastructure.

Lessons of Global Case Studies

The case studies demonstrate that the environmental impact of artificial intelligence infrastructure is deeply interconnected with the environmental conditions of the region, as well as the system of governance . The excessive growth of hyperscale data centers in the United States has caused the issue of water consumption and local control of resources. The strategic planning and strategic thinking on regional resource considerations have been pointed out as crucial in China where the development of digital infrastructure is rapid. In the meantime, European nations have revealed the way in which environmental sustainability policies and positive climatic conditions can contribute to more resource-efficient development of digital infrastructure. A comparative overview of regional data center characteristics and water-sustainability challenges is presented in Table 5.

Table 5. Comparative Environmental Characteristics of Major Data Center Regions

Region	Major Data Center Locations	Key Drivers of Expansion	Climate Conditions	Water Sustainability Challenges	Sustainability Initiatives
United States	Northern Virginia, California, Texas, Arizona	Strong digital infrastructure, high cloud computing demand, presence of major tech firms	Moderate to warm climates in many regions	High water consumption for cooling systems; local water resource pressures	Use of recycled water, water-efficient cooling technologies
China	Beijing, Inner Mongolia, Guizhou, Gansu	Government investment in AI and digital economy, expanding cloud computing sector	Many regions experience arid or semi-arid climate conditions	Data centers located in water-scarce regions	Relocation of data centers to cooler regions; efficiency improvements
Europe	Sweden, Finland, Ireland, Netherlands	Renewable energy availability, strong digital connectivity	Cooler climates suitable for free-air cooling	Lower cooling-related water consumption compared to warmer regions	Renewable energy integration and sustainable data center policies
Middle East	UAE, Saudi Arabia	Digital transformation strategies, economic diversification	Extremely hot and arid climates	High cooling demand increases water consumption	Exploration of seawater cooling and renewable energy integration
Asia-Pacific	Singapore, India, Japan	Rapid digitalization and growing demand for cloud computing services	Tropical and humid climates	Limited land and water resources in some urban regions	Government regulations and sustainability standards for data centers

Source: Author's compilation based on Masanet et al. (2020); IEA (2023); Mytton (2021).

Collectively, these examples reflect the significance of integrating environmental sustainability considerations into digital infrastructure planning and governance. As artificial intelligence technologies continue to expand globally, ensuring that data center development aligns with local environmental conditions and sustainable resource management practices will be vital for minimizing the environmental footprint of the digital economy. These regional and infrastructural dynamics demonstrate that the environmental impacts of artificial intelligence infrastructure extend beyond operational efficiency and require broader discussion concerning sustainability, governance, and environmental justice.

5. DISCUSSION AND IMPLICATIONS

The results discussed in this chapter suggest the multifaceted interconnection between the artificial intelligence infrastructure and the sustainability of freshwater. They increasingly consume resources as the digital infrastructures underpinning artificial intelligence technologies are growing larger, especially hyperscale data centers, in all manners of economic sector. The thematic analysis proves that artificial intelligence creates a water footprint through several interlinked processes, such as cooling systems inside data centers, electricity to run the computational processes, and the production of digital hardware in its lifecycle . All of these processes depict the material bases of digital technologies and refute the widespread view of digital systems that do not rely on environmental resource systems. The water footprint across different stages of the artificial intelligence infrastructure lifecycle is summarized in Table 6.

Table 6. Water Footprint Across the Artificial Intelligence Infrastructure Lifecycle

Stage of AI Infrastructure	Key Activities	Water Consumption Pathway	Environmental Impact
AI Model Development	Training machine learning models using high-performance computing systems	Cooling systems in data centers require freshwater to dissipate heat	Increased water consumption during intensive computational processes
Data Center Operations	Continuous processing, storage, and management of digital data	Evaporative cooling and chilled water systems used to regulate server temperatures	Direct freshwater withdrawals from local water sources
Electricity Generation	Energy supply required to power servers and cooling systems	Thermoelectric power plants consume water for steam generation and cooling	Indirect water footprint associated with energy production
Hardware Manufacturing	Production of processors, GPUs, and semiconductor chips	Semiconductor fabrication requires ultrapure water for wafer cleaning	Water-intensive industrial manufacturing processes

continued on following page

Table 6. Continued

Stage of AI Infrastructure	Key Activities	Water Consumption Pathway	Environmental Impact
Raw Material Extraction	Mining and processing of metals and minerals used in electronic devices	Water used for mineral separation and chemical processing	Environmental stress on water resources in mining regions
Electronic Waste Management	Disposal and recycling of outdated computing hardware	Water used in recycling and material recovery processes	Potential contamination of water systems if improperly managed

Source: Author's compilation based on Hoekstra et al. (2011); Williams et al. (2002); Mytton (2021); Ren et al. (2023).

The water–energy nexus framework provides a useful lens for understanding these dynamics. Data centers require substantial quantities of both electricity and freshwater to sustain computational operations. Electricity generation often depends on water-intensive cooling processes, while cooling systems within data centers directly consume freshwater resources. As AI applications continue to expand, pressure on both water and energy systems is likely to increase. UNESCO (2023) warns that growing industrial and technological demand may intensify global freshwater scarcity, particularly in already vulnerable regions. Scenario-based assessments further suggest that rising computational demand may outpace environmental mitigation efforts, creating additional sustainability challenges (Herrera et al., 2025; D'Orgeval et al., 2026). The lifecycle dimensions of these impacts are summarized, which illustrates environmental pressures across hardware manufacturing, infrastructure construction, model training, deployment, and end-of-life management (See **Figure 4**).

Figure 4. Life Cycle Framework of AI Infrastructure and Associated Enviornmental Impacts

The chapter also contributes to broader discussions on sustainable digitalization and the planetary costs of AI systems (Crawford, 2021). Hlabisa (2025) argues that sustainability assessments should extend beyond carbon accounting to include water use, material extraction, energy consumption, and land occupation. While AI technologies are frequently promoted as tools for climate monitoring, environmental management, and resource optimization, their supporting infrastructures generate significant environmental externalities. Consequently, sustainability considerations must be integrated into infrastructure planning, technology design, and policy development rather than treated as secondary concerns.

A further concern relates to environmental justice. Although the benefits of AI-driven services are globally distributed, the environmental burdens associated with data centers are often concentrated in specific localities. Facilities located in water-stressed regions can intensify pressure on already limited freshwater supplies. The unequal geographical distribution of environmental costs raises important questions regarding resource governance, environmental equity, and community resilience. As digital infrastructure continues to expand, policymakers must ensure that sustainability strategies account for both ecological and social dimensions of development.

The findings also have important implications for policymakers and industry stakeholders. Greater transparency in environmental reporting is necessary to improve understanding of water consumption patterns within AI infrastructure. Technological innovations such as advanced cooling systems, renewable energy integration, and water-efficient operational practices can help reduce environmental impacts. At the same time, long-term sustainability requires strategic planning, including careful site selection, investment in sustainable infrastructure, and regulatory frameworks that promote responsible water management. Ensuring that the continued growth of AI aligns with freshwater sustainability will require coordinated action among governments, industry actors, and research institutions

6. POLICY IMPLICATIONS AND RECOMMENDATIONS

The sustainability challenges associated with artificial intelligence infrastructure require proactive policy interventions that balance technological innovation with environmental responsibility. As hyperscale data centres continue to expand globally, policymakers, industry stakeholders, and regulatory agencies must adopt strategies to minimise freshwater consumption, enhance transparency, and promote sustainable development of digital infrastructure. The following recommendations may support the long-term environmental sustainability of AI systems:

- Enhance transparency by requiring data centres to publicly disclose water and energy consumption data to support environmental accountability and evidence-based policymaking.
- Promote water-efficient technologies, including recycling systems, intelligent water monitoring, and optimised cooling architectures, to reduce freshwater consumption (Hiremath, 2024).
- Encourage the adoption of advanced cooling technologies and renewable energy sources through incentives, sustainability certifications, and targeted funding programs.
- Incorporate environmental impact assessments into data centre planning, particularly in water-stressed regions, to minimise pressures on local freshwater resources.
- Prioritise strategic site selection by locating data centres in areas with cooler climates, renewable energy availability, and adequate water supplies.
- Foster multi-stakeholder collaboration among governments, technology firms, researchers, and international organisations to develop shared sustainability standards and governance frameworks.
- Support the development of standardised metrics and reporting mechanisms for assessing the environmental footprint of artificial intelligence infrastructure.

7. CONCLUSION

The rapid expansion of artificial intelligence (AI) has transformed the digital economy, yet its development remains deeply dependent on material infrastructures and environmental resources. This chapter demonstrates that AI infrastructure, particularly hyperscale data centres, is closely linked to freshwater sustainability through multiple interconnected pathways. First, data centres rely on water-intensive cooling systems to maintain high-performance computing operations. Second, AI-related energy consumption indirectly increases freshwater demand through water-dependent electricity generation processes. Third, the production of digital hardware, including semiconductor manufacturing and raw material extraction, contributes significantly to the overall water footprint of AI systems.

The analysis further highlights how the global expansion of data centres can intensify water stress in regions already facing resource constraints. These environmental impacts vary across geographical contexts depending on climatic conditions, resource availability, and regulatory frameworks. Consequently, sustainable AI development requires a combination of technological innovation, including water-efficient cooling systems and renewable energy integration, alongside stronger governance mechanisms, transparency in environmental reporting, and responsible resource management.

Future research should focus on standardized water-footprint metrics, comparative regional assessments, and lifecycle analyses of AI infrastructure. Ultimately, the sustainability of artificial intelligence cannot be evaluated solely through technological efficiency or economic productivity. Freshwater consumption and broader environmental consequences must become central considerations in future debates on sustainable digitalization and environmental governance.

REFERENCES

Bozkurt, A. (2025). AI's Thirst, AI's Heat, AI's Waste: Exposing the Hidden Environmental Impact of Every Artificial Intelligence Interaction. *Open Praxis*, *17*(4), 638–647. DOI: 10.55982/openpraxis.17.4.1058

Chouksey, A., Rajan, A. K., Gurjar, V., Tiwari, R., & Mishra, P. K. (2026). The green paradox: The climate, environmental, and sustainability implications of artificial intelligence. *Global Environmental Change Advances*, *6*, 100029. DOI: 10.1016/j.gecadv.2025.100029

Crawford, K. (2021). *Atlas of AI: Power, politics, and the planetary costs of artificial intelligence*. Yale University Press.

D'Orgeval, A., Sheehan, S., Avenas, Q., Assoumou, E., & Sessa, V. (2026). Generative AI impact assessment through a life cycle analysis of multiple data center typologies. *Applied Energy*, *406*, 127288. DOI: 10.1016/j.apenergy.2025.127288

de Vries-Gao, A. (2025). The carbon and water footprints of data centers and what this could mean for artificial intelligence. *Patterns (New York, N.Y.)*, *7*(1), 101430. DOI: 10.1016/j.patter.2025.101430 PMID: 41583976

Herrera, M., Xie, X., Menapace, A., Zanfei, A., & Brentan, B. M. (2025). Sustainable AI infrastructure: A scenario-based forecast of water footprint under uncertainty. *Journal of Cleaner Production*, *526*, 146528. DOI: 10.1016/j.jclepro.2025.146528

Hiremath, R. B. (2024). AI-embedded data centres: Promoting sustainability and reducing water footprint. In *2024 First International Conference on Data, Computation and Communication (ICDCC)* (pp. 40–44). IEEE. DOI: 10.1109/ICDCC62744.2024.10961316

Hlabisa, S. (2025). The ecology of artificial intelligence: Energy, water, materials, and land limits of digital systems. *Carbon Neutral Systems.*, *1*(1), 19. DOI: 10.1007/s44438-025-00018-8

Hoekstra, A. Y. (2011a). *The water footprint assessment manual: Setting the global standard*. Routledge.

Hoekstra, A. Y., Chapagain, A. K., Aldaya, M. M., & Mekonnen, M. M. (2011b). *The Water Footprint Assessment Manual: Setting the Global Standard*. Earthscan.

Hoff, H. (2011). *Understanding the nexus: Background paper for the Bonn 2011 conference on the water, energy and food security nexus*. Stockholm Environment Institute.

Jiang, F., Duan, C., & Chen, B. (2025). Facility-level energy-driven water footprint and scarcity implications of Chinese data centers. *Applied Energy*, *399*, 126522. DOI: 10.1016/j.apenergy.2025.126522

Jones, N. (2018). How to stop data centres from gobbling up the world's electricity. *Nature*, *561*(7722), 163–166. DOI: 10.1038/d41586-018-06610-y PMID: 30209383

Lambert, K., & Luccioni, S. (2026). *From Cradle to Cloud: A Life Cycle Review of AI's Environmental Footprint*. ArXiv., DOI: 10.1145/3805689.3812401

Lei, N., Lu, J., Shehabi, A., & Masanet, E. (2025). The water use of data center workloads: A review and assessment of key determinants. *Resources, Conservation and Recycling*, *219*, 108310. DOI: 10.1016/j.resconrec.2025.108310

Li, C., Shu, H., & Gu, X. (2024). Photoperiod Management in Farm Animal Husbandry: A Review. *Animals (Basel)*, *15*(4), 591–604. DOI: 10.3390/ani15040591 PMID: 40003072

Li, P., Yang, J., Islam, M. A., & Ren, S. (2025). Making ai less' thirsty'. *Communications of the ACM*, *68*(7), 54–61. DOI: 10.1145/3724499

Manoharan, G., Ashtikar, S. P., & Nivedha, M. (2024). Unveiling the critical role of artificial intelligence in energy management. In *AI applications for clean energy and sustainability* (pp. 234–253). IGI Global Scientific Publishing., DOI: 10.4018/979-8-3693-6567-0.ch012

Masanet, E., Shehabi, A., Lei, N., Smith, S., & Koomey, J. (2020). Recalibrating global data center energy-use estimates. *Science*, *367*(6481), 984–986. DOI: 10.1126/science.aba3758 PMID: 32108103

Meijer, G. I. (2010). Cooling Energy-Hungry Data Centers. *Science*, *328*(5976), 318–319. Advance online publication. DOI: 10.1126/science.1182769 PMID: 20395502

Mytton, D. (2020). Hiding greenhouse gas emissions in the cloud. *Nature Climate Change*, *10*(8), 701. DOI: 10.1038/s41558-020-0837-6

Mytton, D. (2021). Data centre water consumption. *NPJ Clean Water*, *4*(1), 984–986. DOI: 10.1038/s41545-021-00101-w

Privette, A. P., Barros, A., & Cai, X. (2026). Data Centers Water Footprint: The Need for More Transparency. *AGU Advances, 7*(2), e2025AV002140. DOI: 10.1029/2025AV002140

Ren, S., ... (2023). The water footprint of artificial intelligence. *arXiv preprint*. https://arxiv.org/abs/2304.03271

Samanta, A., & Jiang, Y., [add remaining authors], et al. (2025). Water footprint of datacenter applications: Methodological implications of manufacturing, operational, and decommissioning phases. In *Proceedings of the ACM Symposium on Cloud Computing*. (pp. 102–110). ACM. DOI: 10.1145/3772052.3772216

Strubell, E., Ganesh, A., & McCallum, A. (2019). Energy and policy considerations for deep learning in NLP. *Proceedings of the 57th Annual Meeting of the ACL*. DOI: 10.18653/v1/P19-1355

Vinuesa, R., Azizpour, H., Leite, I., Balaam, M., Dignum, V., Domisch, S., Felländer, A., Langhans, S. D., Tegmark, M., & Fuso Nerini, F. (2020). The role of artificial intelligence in achieving the Sustainable Development Goals. *Nature Communications*, *11*(1), 233–240. DOI: 10.1038/s41467-019-14108-y PMID: 31932590

Williams, E., Ayres, R., & Heller, M. (2002). The 1.7 kilogram microchip: Energy and material use in semiconductor production. *Environmental Science & Technology*, *36*(24), 5504–5510. DOI: 10.1021/es025643o PMID: 12521182

Xiao, T., Nerini, F. F., Matthews, H. D., Tavoni, M., & You, F. (2025). Environmental impact and net-zero pathways for sustainable artificial intelligence servers in the USA. *Nature Sustainability*, *8*(12), 1541–1553. DOI: 10.1038/s41893-025-01681-y

KEY TERMS AND DEFINITIONS

Artificial Intelligence (AI): Computational systems capable of performing tasks that typically require human intelligence, including machine learning, prediction, and decision-making.

Data Centres: Physical facilities that house computing infrastructure, servers, storage systems, and networking equipment used to support digital services and AI applications.

Digital Infrastructure: The interconnected physical and technological systems, including data centres, networks, and cloud-computing facilities, that support digital technologies and services.

Freshwater Sustainability: The responsible management and conservation of freshwater resources to ensure long-term ecological balance and human well-being.

Sustainable AI Infrastructure: The design, operation, and governance of AI-related infrastructure in ways that minimise environmental impacts while supporting technological innovation.

Water Footprint: The total volume of freshwater consumed directly and indirectly throughout the lifecycle of a product, service, or technological system.

Water–Energy Nexus: A framework emphasising the interdependence between water and energy systems, where water is required for energy production and energy is required for water management.

ENDNOTE

1 International Energy Agency. (2023). *Data centres and data transmission networks.* https://www.iea.org/reports/data-centres-and-data-transmission-networks Google. (2024). *Google environmental report 2024.* Microsoft. (2024). *Microsoft environmental sustainability report.* Amazon. (2024). *The Climate Pledge progress report.* Meta. (2024). *Sustainability report.*

Chapter 5
Sustainable Management of Water Resources Amid the Global Expansion of AI Data Centers

K. Ganesh Kumar
http://orcid.org/0000-0002-5719-6193
Velalar College of Engineering and Technology, India

K. Balamurugan
http://orcid.org/0000-0003-4560-3801
K.S.R. College of Engineering, India

T. Sangamithra
http://orcid.org/0009-0009-2090-1571
Velalar College of Engineering and Technology, India

S. Periyasamy
http://orcid.org/0000-0003-0442-0259
Selvam College of Technology, India

N. V. Keerthana
http://orcid.org/0000-0001-8878-2083
Velalar College of Engineering and Technology, India

S. Sujan
http://orcid.org/0009-0005-6237-5021
Velalar College of Engineering and Technology, India

ABSTRACT

The explosive worldwide growth of Artificial Intelligence (AI) infrastructures has added new layers to digital infrastructure complexity while generating an unprecedented demand for natural resources especially water. With increasing computational requirements for AI models, data centers need large cooling systems, many of which rely on substantial amounts of freshwater. This chapter investigates its implications for increasing pressures on global water resources. It analyses evolving water consumption trends in relation to regional vulnerabilities, environmental implications and sustainability issues associated with technological growth. The chapter presents a multi policy intervention-level platform approach to balancing

DOI: 10.4018/979-8-2600-2761-5.ch005

AI innovation and responsible water management (through sustainable cooling technology and strategic site selection). By demonstrating the new possibilities for human-centered AI, the results hope to inspire policymakers, technologists, and environmental planners to cultivate more just and sustainable AI-centric futures.

1. INTRODUCTION

In modern society Artificial intelligence moved towards from experimental research laboratory to infrastructure. AI can power a digital assistants, search engines, medical diagnostic tools, self-driven vehicles, financial analytics, and large-scale language models(Brundage et al., 2018).

The sustainable management of freshwater resources has emerged as one of the most pressing yet underappreciated environmental imperatives in the era of AI infrastructure growth. Computational processes in data centers produce significant heat (Bashroush et al., 2020). All cooling systems therefore need to be such that the operating temperature is safe. However, some cooling systems are water-intensive, especially those based on heat dissipation by evaporative cooling using fresh water(Mytton, 2021). The International Energy Agency (IEA) projects that global data center water demands could reach 100–140 billion m^3 per year by 2030(Medarac et al., 2018)representing a 10–15% increase from 2020 baseline levels, depending on cooling technology deployment and energy grid composition.

As AI workloads scale at an unprecedented rate, the imperative to manage water resources sustainably becomes inseparable from broader goals of environmental stewardship and climate resilience. This chapter synthesizes current knowledge on water sustainability in AI infrastructure, examines regional vulnerabilities and hydrological risks, presents quantitative frameworks for efficiency assessment, and proposes integrated governance mechanisms for sustainable water.

2. AI DATA CENTRES AS ENVIRONMENTAL INFRASTRUCTURE

AI data centers operate as large-scale industrial facilities requiring electricity, cooling infrastructure, water resources, land and construction materials, and network connectivity. It is common for discussions of artificial intelligence data centers to focus mainly on computational power, storage size, and network speed. The environmental effect of these facilities is to serve as large-scale industrial systems that continuously process natural resources. Unlike the environment for traditional computing, processing is concentrated in modern AI data centers, which consume a considerable amount of energy and generate a large amount of heat(Masanet &

Lei, 2020). The management of this heat output involves intricate cooling systems that rely not just on electricity but also on water supplies.

In this regard, one should not look at AI data centers as only digital infrastructure. They should be viewed as environmental infrastructure on the scale of a power plant, a factory, or a waste water processing plant. They require a steady flow of resources, produce waste heat, pollute the air, and strain the environment, and all of this must be managed sensibly if they are ever to be sustainable.

The growth in scale for AI workloads also changes in global systems of resources. Previous generations of data centers were designed primarily for storage and rudimentary computing services. By contrast, they are designed to support deep learning models and real-time analytics as well as cloud-based AI services with much greater computational intensity. Consequently, the ecological impact of the data center had increased dramatically during the past ten years.

Tight binding between energy and water is one of the key features of AI data centers. High performance computing (HPC) processors produce a significant amount of heat during their operation and this heat should be dissipated very efficiently to avoid hardware failure. Hence, the cooling equipment is a vital element to the system(Haghdadi et al., 2018). While air-based cooling technologies were adequate for earlier computing architectures, today's AI servers tend to necessitate liquid or evaporative cooling solutions that are known for being water-intensive.

This dynamic of computing, heat generation and cooling need contributes to what is known as the water–energy nexus in digital infrastructure(Zhang et al., 2020). More computation consumes more electricity, and more electricity produces heat. To dissipate this heat, water is used in cooling systems, particularly in evaporative cooling towers. As such, the development of artificial intelligence indirectly adds more strain to water sources.

Another factor that makes AI data centers so environmentally significant is that they never stop running. But data centers are more like factories than mines: they run continuously, day and night. Resource consumption is also continuous, a by-product of this constant vitality. Even minor inefficiencies in cooling or water use, can add up to large environmental effects over time.

Climate also significantly affects the impact on the environment within data centers. Many are located in areas of the country with low electricity rates or with good network connectivity. However, such sites do not necessarily have sufficient water to support large-scale cooling.

Data centers also indirectly use water when generating the electricity they consume (Gehring & Cand, 2024). Thermoelectric power plants, which many and can be majority of electricity generation worldwide, are also water-intensive for cooling and steam. Thus, the water footprint of AI infrastructure encompasses not only the water used in a data center, but also the water used to supply the electricity needed to run the data center. This indirect usage deepens the intertwined relationship between digital technology and environmental sustainability.

Table 1. AI Data Centers as Environmental Infrastructure

Component	Role in Data Center	Environmental Resource Used	Environmental Impact	Water Sustainability Implication
Electricity	Powers servers, storage, and networking	Energy (grid / power plants)	High energy demand, indirect water use, carbon emissions	Coal-heavy grids add 2.0–3.5 L/kWh indirect water footprint
Cooling Systems	Removes heat from processors	Water, electricity, cooling chemicals	Water consumption, heat discharge	WUE ranges from 0.02 to 3.0 L/kWh by technology
Water Resources	Used in evaporative / liquid cooling	Freshwater supply	Water depletion, local resource stress	Sustainable sourcing, recycled water, and closed-loop systems
Land & Construction	Buildings, hardware facilities	Land, steel, concrete, rare materials	Land use, material extraction impact	Site selection in low water-stress regions reduces long-term demand conflict
Network Infrastructure	Data transfer and connectivity	Energy, cables, electronics	Additional energy consumption	Indirect contribution to grid-based water demand
Continuous Operation	24/7 processing	All above resources continuously	Long-term environmental load	Persistent water withdrawal creates chronic stress in water-scarce regions

In Table 1 shows the major infrastructure components required for AI data centers and their associated environmental impacts.

Therefore, modern day AI data centers should be conceptualized as nodes in a regional or global environmental system that includes energy production, water resources, climate conditions, and ecological equilibrium. Overseeing such a system entails integrated planning for technical efficiency and for environmental constraints.

3. DIRECT WATER USAGE EFFECTIVENESS (WUE) IN DATA CENTER COOLING

Water use is one of the most challenging environmental impacts of today's data centers, especially those built for artificial intelligence workloads. Sustainable water management in AI data centers begins with accurately quantifying and minimizing direct water consumption. With increasing density, the processors are producing a lot of heat that have to be removed continuously to keep a stable operation. Consequently, ventilation systems have increasingly become a critical infrastructure requirement, and in many cases, these systems have leaned heavily on water-based solutions. Increased dependence on these cooling techniques has led to water consumption becoming a significant variable in determining the sustainability of digital infrastructure.

High water use in data centers is the result of the thermal energy that high-performance computing equipment produces. For AI workloads in particular, such as deep learning and large language models, specialized processors like GPUs (graphics processing units) and TPUs operating at very high-power levels are required. These generate far more heat than traditional servers, making standard air-cooling systems inadequate. Many data centres, in fact, implement evaporative cooling – or liquid cooling solutions that use water to draw heat from the hardware.

Inside the facility, the water changes from a liquid to a gas and in so doing absorbs heat, the facility temperature is kept at safe levels. This process of cooling is energy efficient when compared to refrigeration, but a constant supply of freshwater is required. Evaporative cooling systems, where water transitions from liquid to gas and absorbs heat, require continuous freshwater supply, with consumption rates typically ranging from 1–3 L/kWh. Depending on the size of the data center and the intensity of the computational load, the quantity of water used by large cooling towers can be enormous. When you multiply that consumption by hundreds of facilities around the globe you have a massive water demand.

Liquid cooling is another popular cooling technique, which involves pumping coolant directly through or very close to the electronics components. While this process can be more efficient than air cooling, in many applications it still needs water – either via the coolant system or as a tertiary cooling step. Certain advanced data centers will employ closed-loop systems to recycle water. However, such systems still need to be replenished periodically because of evaporation, leakage, or other CMM processes. Figure 1 represents the water consumption of data centers.

Figure 1. Water Consumption in Cooling systems modern AI data centers

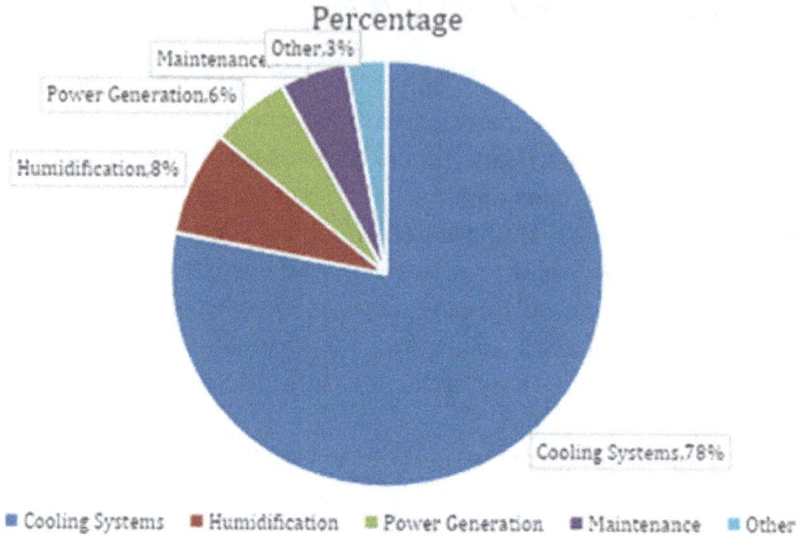

The use of water in data centers is affected by a variety of elements, such as weather conditions, cooling methods, workload density, and site design. Facilities situated in warm environments need to be cooled more heavily than those in cool environments to stay within acceptable temperature limits, leading to higher water consumption. Likewise, data centers with a significant portion of their workload devoted to large-scale AI training may be more water-intensive than sites used mainly for storage or simple computing work. These differences make it impossible to define a single global value for water intensity. However, the literatures support that majority of water used in a data center is associated with cooling systems.

Apart from the direct water consumption, there is also water use — indirect water use — related to production of electricity. Many types of power plants, including thermal and nuclear, use water for cooling and steam generation. Since data centers are energy-hogging, AI infrastructure's overall water footprint includes not just the water used to cool the servers onsite, but that used to produce the electricity powering those servers. This indirect use is often referred to as virtual water, and can be just as significant of the direct water use of the data center itself.

The energy-water nexus alone makes the case for looking at digital infrastructure through a lifecycle lens. Enhancing the efficiency of cooling can reduce water use, but it might raise energy use if other technologies consume more electricity. On the other hand, energy saving techniques may call for more water intensive cooling. Sustainable design, then, must walk a very fine line between these competing issues.

When assessing water use for data centers, scientists typically cite performance metrics such as Water Usage Effectiveness (WUE) that expresses the volume of water consumed against the amount of energy used by the computing hardware. Lower WUE numbers are better, meaning less water is used. This measure has become a valuable metric for evaluating various cooling technologies and steering the advancement of more sustainable facilities.

Water use impact on the environment also varies based on the water availability in the region in which the data center is situated(du Plessis, 2023). The effect of refrigeration machines could be slight in countries where water is extremely plentiful. But in drought-stricken or water-scarce areas, even small amounts of water use can cause large-scale environmental and social turmoil. That's why many experts say water accessibility should factor in from the start when planning new data centers. Over the past few years, some technology companies have started investigating different cooling techniques that are less reliant on fresh water. These include seawater cooling, air-side economizers, immersion cooling and advanced liquid cooling solutions that more efficiently recapture water.

Data centers use different cooling technologies depending on computational load, climate conditions, and infrastructure design. Each cooling method has a different level of water consumption and efficiency. Traditional air cooling requires less water but may not be sufficient for high-performance AI workloads, while evaporative and liquid cooling systems provide higher efficiency at the cost of increased water usage. The comparison of commonly used cooling technologies is shown in Table 2.

Table 2. Cooling Technologies used in Data Centers

Cooling Technology	WUE Range(L/kWh)	Efficiency	Adoption Level	Sustainability Status
Air Cooling	0.0–0.5	Moderate	Legacy systems, limited AI workload	Acceptable
Evaporative Cooling	1.0–3.0	High energy efficiency	Google (0.2–1.5 L/kWh)	Unsustainable in water-stressed regions
Liquid Cooling	0.5–1.5	Very high efficiency	Microsoft (emerging) growing adoption	Improving; transition pathway
Closed-loop Systems	0.1–0.5	Extremely efficient	Specialized installations, increasing deployment	Sustainable

continued on following page

Table 2. Continued

Cooling Technology	WUE Range(L/kWh)	Efficiency	Adoption Level	Sustainability Status
Immersion Cooling	0.02–0.15	Extremely efficient	AWS (<0.5 L/kWh) emerging technology	Best-in-class sustainable
Seawater Cooling	0.1–0.4	High efficiency	Scandinavian facilities.	Sustainable where available

As shown in Table 2, evaporative cooling systems are widely used because they provide high thermal efficiency, but they also require significant amounts of water. Immersion cooling and closed-loop systems offer better water efficiency, although their adoption is still limited due to higher installation cost and technical complexity. The selection of cooling technology therefore plays a major role in determining the environmental impact of AI data centers.

4. INDIRECT WATER FOOTPRINT: GRID-SPECIFIC ANALYSIS

The total of indirect water use represents the amount of water used to produce electricity. Water usage from the grid is influenced by how many different types of electric generators are used in that area. Thus, it would seem that a full approach toward developing a long-term water management plan for data centers using artificial intelligence should include all forms of water consumed by these facilities.

The formula for finding indirect water consumption is:

$$WFindirect = (EIT) * (PUE) * (WFgrid)$$

Where,

EIT = the average kWh per year,
PUE = Power Usage Effectiveness,
WFgrid = water usage (in liters) of each type of electric generator per kilowatt hour produced.

There is considerable variability in the typical water intensity for various grids, Coal-fired electric generating stations have a range of 2.0 – 3.5 liters/kilowatt-hour, Nuclear electric generating stations have an average range of 1.5 – 2.5 liters/kWh, Hydroelectric generating stations produce essentially no water loss (from 0.00 – 0.20 L/kWh). Wind and Solar electrical production result is(less than .050 L/kWh). Therefore, when comparing facilities located in areas with coal-generated electricity versus those operating on renewable resources, facilities dependent upon coal generated electricity will experience increased overall water at rates ranging from approximately 4 – 5 L/kWh.

Table 3. Indirect Water Footprint Analysis: Grid Composition and Calculation Examples

Power Grid Type	Water Intensity (L/kWh)	Regional Examples	Indirect WF per Facility (L/kWh at PUE=1.67)	Sustainability Rating
Coal-based (>60%)	2.0–3.5	US (southwest), India (Bangalore, Pune), Poland	3.3–5.8	Critically unsustainable
Natural Gas (primary)	0.5–1.0	California, Turkey, Middle East	0.8–1.7	Moderate concern
Nuclear (>40%)	1.5–2.5	France, Ontario, South Korea	2.5–4.2	High concern
Hydroelectric (>50%)	0.0–0.5	Norway, Iceland	0.0–0.8	Sustainable
Renewable mix (>70%)	0.0–0.2	Denmark, Costa Rica, Uruguay	0.0–0.3	Highly sustainable
Mixed grid (typical)	0.8–1.5	US national average, EU average	1.3–2.5	Improvement needed

The above table 3 states that the indirect water footprint analysis by composition grid and calculations.

5. GLOBAL DISTRIBUTION OF DATA CENTERS AND WATER STRESS

The worldwide surge in artificial intelligence has resulted in a dramatic increase in the number and scale of data centres globally. Such facilities are not randomly scattered, their location decisions are influenced by economic, technological and environmental considerations including electricity cost, network connectivity, availability of land and climate conditions. But the environmental consequences of those decisions of where to site remain are increasingly relevant – especially with respect to water. As modern data center cooling systems often rely on freshwater, the spatial distribution of these centers can influence regional water resources.

Hyperscale data centers have been built in many parts of North America, Europe and Asia in recent years to provide cloud computing and artificial intelligence services(da Silva et al., 2025). Sustainable water resource management requires explicit consideration of regional hydrological conditions in data center siting and operational decisions. These massive complexes must be cooled continuously to keep high-performance processors operating at safe temperatures. Consequently, their water usage can be significant, particularly in plants employing evaporative cooling towers. If several data centers are clustered in a single area, their aggregate water needs may strain local supplies, especially in regions with existing water shortages.

Availability of cheap electricity is one of the major reasons why data centers are placed in certain areas. AI workloads are energy intensive, and operators tend to pick sites where power is cheap and reliable. But areas where energy is cheap are not always areas with abundant water. Drought feeders Some data centres have nevertheless been located in semi-arid or drought subject areas on the basis that power needs are great, and while their water supplies are small. This discrepancy between energy and water availability in the transboundary context may challenge the sustainable development of environment.

Climate conditions also significantly impact cooling needs. Data centers in warmer climates need more cooling to keep their operating temperatures stable and use more energy and water. By contrast, centers established in colder climates need to use outside air to cool the centers reducing dependency on water-based systems. That's why several tech firms are now locating data centers in colder climes to boost efficiency and cut environmental damage. Still, the demand for digital services is growing worldwide, and new data centres are being built in a variety of locations, even in areas where water is already scarce.

Data centers are often located based on economic factors such as energy cost, tax incentives, and network infrastructure. Environmental sustainability considerations have historically received less attention.

The global spread of data centers is also influenced by the availability of high-speed network infrastructure. Applications for artificial intelligence (AI) need data to move quickly, which can restrict location options to places with modern communications infrastructure. Thus, many data centers are established close to large cities or industrial zones, where the water need of other industries is already significant(Organization et al., 2022). In these areas, large-scale additional cooling systems may exacerbate competition for water with industry, agriculture and the population that's the only thing we really fear.

The environmental Further, impact of water consumption in a data center relies on more than just how much water is consumed. In areas of high rainfall and well-developed water management, the effect may be minimal. Yet in regions characterized by drought, depletion of groundwater, or the effects of climate, even modest water consumption can lead to serious repercussions. For instance, some areas have seen groundwater levels drop partially due to industrial water use – like cooling systems in massive buildings. Figure 2 Represents the usage of water in data centers and the usage in future.

Figure 2. Illustrates the water usage of AI data centers throughout the year and prediction of future use

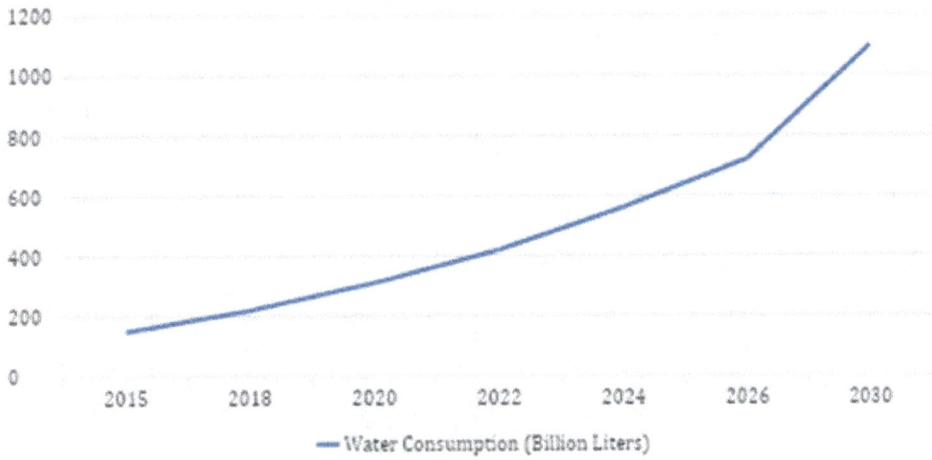

As a result of these concerns, the sustainable viability of data center siting has begun to be assessed by water risk metrics for researchers and government officials. These analyses take into account regional water availability, climate change, population demand, and industrial use. By balancing water needs and availability, we can predict if a new data center site will have a low potential for environmental impact. Such assessments are critical as AI infrastructure continues to grow.

Beyond the direct water use, the global distribution of data centers also has an influence on the indirect water consumption associated with electric power generation. Numerous areas depend on thermal power plants, which need water for cooling and to make steam. Data centers consume huge amounts of electricity, and by doing so they indirectly increase water demand at power plants. This usually impacts the water consumption of power, that is why we often talk about water footprint of electricity, and it is significant in places using a lot of water-intensive power.

In order to find out what location and sustainability is related to, it helps to see how the region stacks up in relation to both data center growth and the water stress indicator. The table below lists typical conditions in various regions of the world.

Table 4. Global Data Center Expansion by Region

Region	WSI*	Growth Rate	Primary Cooling Technology	Mitigation Status	Sustainable Management Priority
US Southwest (AZ, NV, UT)	0.8–1.0 (High)	15–20%/yr	Evaporative cooling dominant	Some facilities adopting dry cooling	Critical – immediate transition required
India (Bangalore, Pune)	0.7–0.9 (High)	25–30%/yr	Hybrid systems emerging	Water recycling targets <30% adoption	Critical – policy intervention urgent
Ireland (Dublin, Cork)	0.4–0.5 (Moderate)	10–12%/yr	Mixed air/liquid cooling	Regional water policies emerging	High – proactive governance needed
Netherlands (Amsterdam)	0.3–0.4 (Moderate)	8–10%/yr	Liquid cooling systems	Advanced water management protocols	Moderate – continue innovation
Singapore	0.9–1.0 (High)	Limited expansion	Liquid cooling only	Water constraint-driven policy	Critical – moratorium enforced
Northern Europe (Norway, Sweden)	0.1–0.3 (Low)	5–8%/yr	Air-side economizers	Climate-optimal; minimal intervention needed	Low – preferred expansion zone

The table 4. indicates that the majority of data center growth strongholds are also areas with limited or overused water supplies all over the world. This is a situation that underscores the necessity of thoughtful planning in terms of new AI infrastructure siting.

Various measures can be taken to mitigate the environmental risk. Such measures involve the choice of site in a water-rich area, the utilization of less-water-intensive cooling technologies, the re-use of water in the plant, and the adoption of energy-efficient processes to reduce heat generation. Besides, there could be new legislation at the government level requiring environmental reviews before new data centers can be built.

With the rapid growth of worldwide artificial intelligence, the link between data center location and water sustainability will gain even more prominence. Future infrastructure development planning must take into account not only the technological capabilities and economic costs of the supplies, but also the availability of natural resources for a long period of time.

Thus, the geographic distribution of data centers around the world is important when considering the impact of artificial intelligence on the environment. Cooling, water use, and indirectly through electricity production, all are affected by site choice. With integrated water risk management at the infrastructure level, the continued success of AI technologies can be enabled with significant protection of vital environmental resources.

6. MATHEMATICAL FRAMEWORKS FOR WATER EFFICIENCY ANALYSIS

Rigorous sustainable water management in AI data centers requires quantitative frameworks enabling precise efficiency assessment, technology comparison, and environmental risk prediction.

Although the qualitative narrative is informative to help understand the general trends, mathematical frameworks allow a more rigorous analysis of how water is used and can be saved. These models provide the ability to predict water demand, assess cooling performance, and determine sustainable infrastructure configurations.

Energy consumption, heat generation, and cooling are all intricately linked in modern data centers, which are highly dynamic environments. As more heat is generated by growing computational demands, more cooling is necessary. In many are such facilities that use water-based systems to accomplish this cooling, it is important to analyse the connection between computing power and water consumption. A number of performance indicators have been proposed to quantify this relation, such as the WUE (Water Usage Efficiency), the water to energy efficiency metrics, the evaporation loss models, and the regional water risk indices. They help decision makers to evaluate multiple cooling options, to choose appropriate data center locations, and to design infrastructure that reduces environmental impacts.

Figure 3. Illustrates the strong relationship between energy consumption and water usage in AI data centers

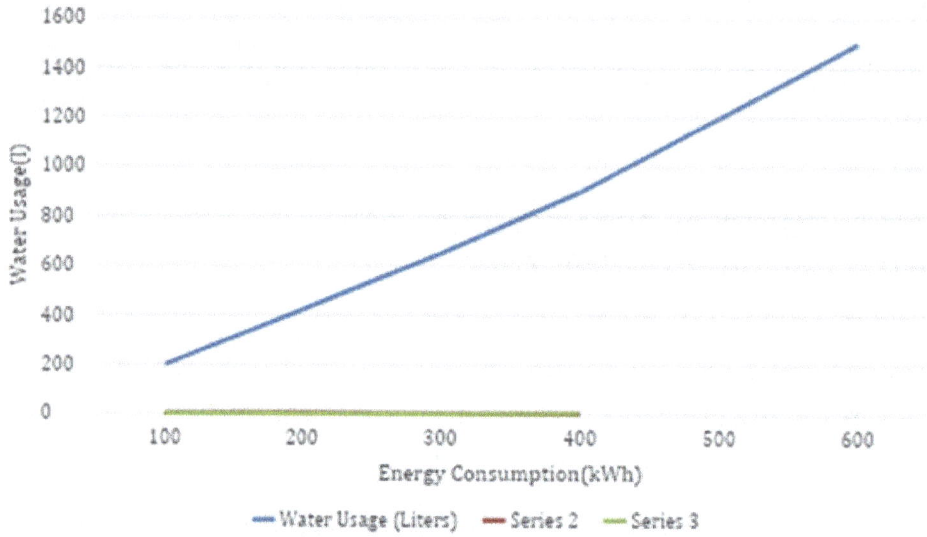

6.1 Water Usage Effectiveness (WUE)

An industry standard for water efficiency in data centers is the Water Usage Effectiveness (WUE) metric. This measurement quantifies the water usage associated with the energy used by information technology equipment.

Formula

$$WUE = \frac{Annual\ Water\ Usage}{IT\ Equipment\ Energy}$$

Where:

Water Usage = Total water consumed annually (liters)
IT Equipment Energy = Total energy used by computing equipment (kWh)

Example

A facility consuming 1,200,000 L/year with 600,000 kWh/year IT energy.
Assume that a data center consumes:
Water Usage = 1,200,000 Liters/year

IT Energy = 600,000 kWh/year

$$WUE = \frac{1{,}200{,}000}{600{,}000} = 2.0\ L/kWh$$

The data center uses 2.0 Liters of water for each kilowatt-hour of computing energy.

Lower WUE values indicate better water efficiency. Sustainable management targets set WUE < 0.5 L/kWh for new facilities by 2027 (IEA, 2023).

6.2 Water–Energy Coupling Efficiency Model

Water use and energy use are tightly coupled in data centers. Energy-efficient cooling systems can consume more water, and water-saving systems typically consume more energy. To assess this trade-off, a water–energy efficiency ratio can be defined. Sustainable management strategies should maximize WEER while minimizing absolute WUE.

Cooling systems create a relationship between water consumption and energy efficiency.

Water–Energy Efficiency Ratio (WEER)

$$WEER = \frac{Cooling\ Energy\ Reduction}{Water\ Consumption}$$

Where:

Cooling Energy Reduction = Energy saved by using water-based cooling
Water Consumption = Water required for cooling

Higher WEER values indicate more efficient water–energy trade-offs.

Example

Energy saved = 100,000 kWh
Water Consumption = 40,000 Liters

$$WEER = \frac{100000}{40000} = 2.5$$

The cooling system saves 2.5 units of energy for each unit of water consumed.

This model is useful when comparing evaporative cooling with air cooling or liquid cooling.

6.3 Cooling System Water Loss Model

Evaporative coolers extract heat through the evaporation of water. How much water is needed is a function of the heat load of the servers. The evaporation loss can be calculated from thermodynamics.

Water loss in evaporative cooling systems can be estimated using thermodynamic heat removal calculations. This model informs facility-level water budgeting and identifies thermodynamic limits of evaporative system optimization.

Water Evaporation Model

$$W = \frac{Q}{h_{vap}}$$

Where:

W= Water evaporated (kg)
Q= Heat removed (kJ)
h_{vap}= Latent heat of vaporization (~2257 kJ/kg)

Example:
Heat load = 4,514,000 kJ

$$W = \frac{4{,}514{,}000}{2257} \approx 2000\,kg$$

Approximately **2000 Liters of water** are required.

6.4 Regional Water Risk Index (RWRI)

But beyond efficiency, you also need to assess whether the data center is situated in an area that has abundant water supplies. A straightforward index can be applied to assess environmental risk.

Sustainable siting decisions require assessment of local water availability. The Regional Water Risk Index quantifies facility-level risk:

RWRI Model

$$RWRI = \frac{W_d}{W_a} \times C_f$$

Where:

W_d= Water demand from the data center
W_a= Available regional water supply
C_f= Climate stress factor(1.0–2.5).

RWRI > 0.3 indicates potential environmental concern requiring mitigation. This index should be a mandatory component of environmental impact assessments for new data center developments in water-stress regions.

7. TECHNOLOGICAL INNOVATIONS IN WATER-EFFICIENT COOLING

Technological advancement is the most direct pathway to sustainable water management in AI data centers. Growth in computing requirements for artificial intelligence has made cooling efficiency among the most critical issues in the design of current data centers. Traditional cooling accounts for a considerable proportion of the total water consumption, especially in plants that use evaporative cooling systems, as commented before. Given the constraints on water availability worldwide, and the increasing demand for water, there is a rising interest in the development of new cooling technologies that are capable of lowering water use without compromising performance. So, advancing technology is key in making sure the expansion of AI infrastructure is not only financially but environmentally sustainable.

Due to the presence of high-performance processors (graphics processing units (GPUs), tensor processing units (TPUs) and specialized AI accelerators), modern data centers produce very high levels of heat. These components are run at high power densities and must be cooled continuously to prevent overheating. Traditional air-cooling solutions don't work well for these kinds of loads, which is why water cooling is so common. Nevertheless, the energy impact of such strategies spurred researchers and engineers to seek solutions that would increase both energy efficiency and water efficiency.

Among the most exciting new technologies is liquid cooling, in which the coolant flows directly over or very near the electronic parts, rather than cooling air that is circulated around the component. Since liquids are better conductors of heat than air, liquid cooling solutions can dissipate heat with fewer energy inputs and, in certain instances, less water. Direct-to-chip cooling, for instance, permits coolant to pass through tiny channels that are connected to chips, diminishing the need for vast cooling towers. What's more, this process reduces the evaporation loss accompanied by conventional cooling schemes.

The other is immersion cooling, where servers are submerged in a bath of non-conductive liquid. The liquid takes up heat from electronic parts directly, and then delivers it to an ex-situ cooling mechanism. Immersion cooling can greatly lower the amount of evaporating water required, as large volumes are not required to bathe components in water. Plus, immersion cooling enables servers to be "run at much higher temperatures without damage — further lessening the demand for aggressive cooling." While this technology is still being phased in and is among the most efficient solutions for future AI data centers.

Closed loop cooling systems are also an option for water conservation(Vaughn, 2021). Water is recirculated in these operations within the plant rather than being released into the environment after one time use. Heat exchangers transfer thermal energy without evaporating water. You still have to make up for some losses, but the overall usage is a fraction of what it is in an open evaporative system. Closed-loop systems are particularly conducive to operation in water-constrained areas as they impose minimal withdrawal rates from local aquifers.

Some plants, particularly in coastal areas, have begun to utilize seawater cooling in place of fresh water. Given that seawater is readily available in vast quantities in many places, it makes sense to use it to dissipate heat rather than to continue to strain scarce fresh water supplies. Seawater is pumped through heat exchangers that extract heat from the data center without coming into contact with internal cooling system. This method involves an exchange of heat but not of fluids, which is to say the seawater does not get in contact with the internal cooling systems. While specialized machinery needs to be designed to protect against corrosion, it is a feasible approach for large plants situated by the ocean.

AI-driven thermal management systems represent a particularly promising innovation for sustainable water stewardship: these systems dynamically optimize cooling operations based on real-time temperature, workload, and environmental data, reducing overcooling and energy–water coupling losses by up to 30%(Alsouda et al., 2023)

Figure 4. compares the efficiency of different cooling technologies used in modern AI data centers.

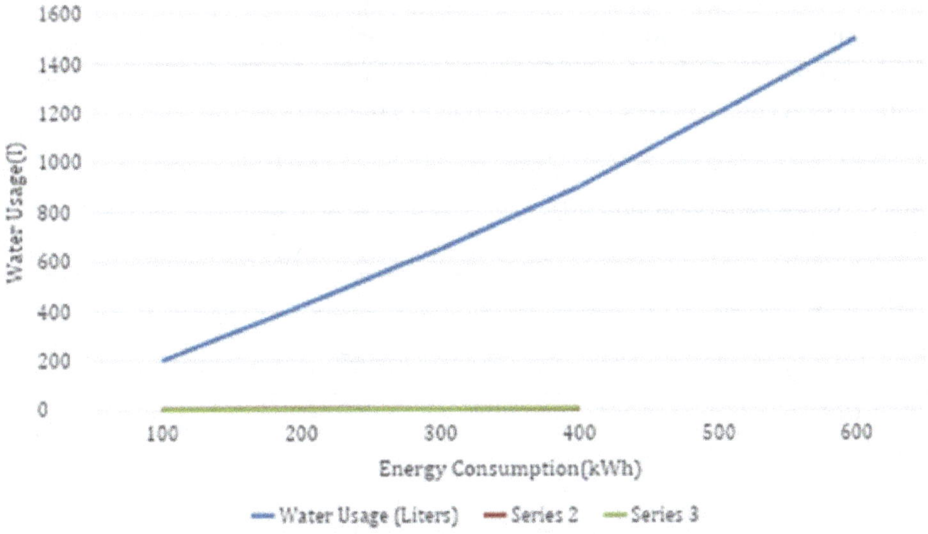

There's also another technology that's been in the news lately: air-side economizers, which enable data centers to employ outside air for cooling when environmental conditions permit. Outside air can be directly used to cool the interior of the building in colder areas, negating the need to use water-based cooling for part of the year. This approach slashes energy and water use by a significant amount. Consequently, many firms now are constructing data centers in colder climates where outside air can be employed for natural cooling much of the year.

And cooling efficiency is being improved through the use of artificial intelligence itself. Thermal management solutions based on AI are able to gather information such as temperature, workload, and environmental factors in real time, and dynamically control the cooling operations by themselves to the environment. By balancing the airflow, coolant flow, and the distribution of workload among servers, these systems are able to eliminate overcooling which leads to reducing water consumption to a minimum. Machine learning (ML)-based algorithms can foresee rising cooling demand and the system adjusts ahead of potential excess heat generation, which enhances both system performance and energy efficiency.

In fact, the layout of the data center building can also contribute aiding in cooling efficiency, not just improvements in technology. Contemporary buildings and facilities often incorporate optimized airflow paths, advanced insulation techniques, and modular construction methods that enable heat to be more efficiently removed. Certain data centers are constructed underground or underwater to utilize natural temperature stability, thus minimizing use of active cooling. These design breakthroughs prove that new equipment isn't the only way to boost sustainability and that the building itself can be more energy- and water-efficient.

To contrast various cooling technologies, it is helpful to examine their water use, efficiency, cost, and maturity. The attributes of several cooling techniques widely employed in AI data centers are summarized in the table below.

Table 5. Water Reduction Potential

Technology	Water Reduction
Immersion Cooling	80–95%
Direct Liquid Cooling	60–80%
Closed-loop Cooling	50–70%
Seawater Cooling	Up to 90%
AI Thermal Optimization	20–40%

In this table 5. it illustrates the technologies show strong potential for sustainable infrastructure development.

8. POLICY AND GOVERNANCE FRAMEWORKS

The buildout of artificial intelligence infrastructure is so rapidly growing that it is spawning challenges for which technology solutions will not really suffice. The long-term sustainability of data center water and energy use is influenced not only by design and technology (e.g. efficiency improvements in cooling systems, information technology, and energy supply) but also by policy and governance. Governments, regulatory bodies, and international institutions have an essential role to play in ensuring that such growth in digital infrastructure does not surpass the environmental capacity of the regions where it is domiciled. In the absence of appropriate regulations and foresight, the heightened demand for water for data center cooling could exacerbate resource conflicts, particularly in regions most vulnerable to water scarcity.

Data center sustainability-related policies generally have three focus areas: resource management, environmental sustainability, and infrastructure development. Resource management legislation aims to limit the water and energy consumption of industrial plants, such as major data centers. Environmental conservation regulations seek to minimize detrimental influence of cooling on wildlife, ground water, climate patterns, and so on. Infrastructure planning policies are also in place to confirm that new data centers are constructed in areas where they can have access to resources but aren't a detriment to surrounding communities or ecosystems.

Among the key policy levers deployed recently is the obligation for an environmental impact assessment (EIA) prior to the establishment of a large data center. Environmental impact assessments predict the impact of projects on surface water, ground water, air and land, among other areas(Silander, 2019). Governments can assess in advance whether a facility can be allowed to operate without affecting the environment when they consider these elements. In certain nations, new data center approval is only given if the operator can prove its water usage will be sustainable.

Another key policy instrument is the introduction of water use reporting and water use transparency. Companies are required to quantify and report how much water they are using at their plants under these rules. This enables regulators and scientists to track the environmental consequences of digital infrastructure and to pinpoint where the greatest gains in efficiency can be made. And public disclosure also pressures companies to become more sustainable, since their environmental performance ends up as part of the company's corporate responsibility commitments.

Governments may also set water prices and allocate water to avoid over use. In water constrained areas, industrial users may face higher prices for large quantities of water which may provide them with an incentive to use the most efficient cooling technology. In certain instances, there also are restrictions on how much water a plant can pull from local sources. This is done to minimize the risk of industrial demand for water displacing supply for agriculture, households or natural ecosystems.

Regulations on site selections are also a crucial element of governance. Since the environmental impact of a data center is highly location-dependent, some nations have enacted regulations that prohibit building large data centers in environmentally water-scarce regions. Rather, they should focus on building new data centers in regions that have drinking water to spare and where the natural climate is conducive to cooling. For example, cooler temperatures can diminish evaporative cooling needs, and seawater can be used for heat dissipation in coastal areas. In their guidance for infrastructure development in suitable location, policymakers can help to mitigate the environmental impact of domestic AI proliferation in the future.

International coordination will become increasingly necessary as the data center industry expands worldwide. Artificial intelligence services are frequently run by large international companies, with their data centers possibly being situated in several different countries. Since water availability and environmental legislation differ greatly from one region to another, international guidance can provide a basis for common standards of sustainability. Bodies like the International Energy Agency (IEA), the United Nations Environment Programme (UNEP) and the European Commission, among others, are starting to release guidance on energy efficiency, water management and environmental disclosure for the digital infrastructure(Clarke, 2025). These recommendations also nudge companies to do their best, even in the face of lenient local laws.

In addition to this, many tech companies have their own sustainability policies. These corporate commitments typically include objectives such as reducing water usage, utilizing recycled water, increasing energy efficiency and becoming carbon neutral. Some businesses have committed to being water positive, returning more water to the environment that they use. This can be done through water recycling, conservation efforts or support of local water management initiatives. While these are voluntary programs, they illustrate that industry engagement in promoting best practices for sustainable and low carbon can be effective in achieving improvements.

Effective sustainable management of water resources in the AI data center sector requires integrated governance involving utilities, data center operators, regulators, and communities. When considering the efficiency of various policy mechanisms, a comparison of the primary governance instruments in data center sustainability management can be informative.

Table 6. Multi-Stakeholder Governance Framework for Sustainable Water Management in AI Data Centers

Stakeholder	Primary Role	WUE Performance Target	Permitting/ Compliance Timeline	Enforcement Mechanism	Water Mgmt. Priority
Data Center Operators	Technology adoption; facility design	<0.5 L/ kWh for new facilities by 2027	Annual reporting; 3-year compliance reviews	Performance penalties, water extraction restrictions	High
Utilities / Water Authority	Water resource allocation; infrastructure planning	Maintain basin-level WSI <0.6	Regional water stress assessments (biennial)	Extraction caps, tiered pricing escalation	Critical
Regulators / Government	Policy development; environmental protection	Sector average reduction to <1.2 L/kWh by 2030	Facility siting review (6–12 months); EIA mandatory	Permit denial, emission taxes,	Critical
Community / Environmental Groups	Impact monitoring; transparency advocacy	Transparent, publicly available WUE data	Quarterly public reporting requirements	community veto authority in water-scarce regions	High

The Table 6. illustrates that Multi-Stakeholder Governance Framework for Sustainable Water Management in AI Data Centers.

With the advances in AI, particularly, in governance and policy are going to matter more and more. In the future, data centers will have to comply with more stringent environmental rules, and proof will be required that their infrastructures can be run without exerting too much pressure on Mother Nature. Existing laws might also need to be amended to sufficiently regulate the new issues arising from HPCS and large-scale AI systems.

In the final analysis, technology advancement is not enough to make modern data centers sustainable. Policy and regulatory reform that balances the potential benefits of AI with the imperative to safeguard global hydric resources is also needed. Through environmental regulations, well-planned infrastructure, and responsible business behaviour, it is feasible to nurture the growth of AI and long-term ecological health simultaneously.

Recommended policy instruments for sustainable water management include: (1) mandatory WUE disclosure in environmental impact assessments for all new data center developments; (2) water stress index-based siting restrictions prohibiting evaporative cooling in regions with WSI > 0.7; (3) performance-based incentives rewarding sub-0.5 L/kWh achievement; (4) integrated water-energy planning frameworks at the regional level; and (5) cross-border coordination mechanisms for transboundary water systems supporting shared AI infrastructure hubs.

9. ENVIRONMENTAL AND HYDROLOGICAL IMPACTS

Water-intensive data center operations create cascading environmental consequences: aquifer depletion, river flow reduction, thermal pollution, biodiversity disruption, and agriculture–industry–residential competition for limited water supplies. Climate change exacerbates these impacts through intensified drought cycles and increased evaporative cooling demands. Sustainable water management frameworks must therefore incorporate climate adaptation planning, including scenario-based stress testing of facility water supply chains under projected climate trajectories (du Plessis, 2023). Mitigation requires site selection informed by hydrological indices, technology innovation, and participatory governance mechanisms that include affected communities.

10. LIFECYCLE WATER ANALYSIS: EMBODIED VS. OPERATIONAL WATER

Complete sustainable water management accounting must encompass embodied water in hardware fabrication, not only operational consumption. GPU production requires approximately 10–20 L of water per kWh of equivalent computing capacity over its operational lifetime. When integrated with operational water consumption, lifecycle water footprints substantially exceed direct-use estimates, underscoring the importance of hardware longevity, recycling, and efficiency optimization.Procurement policies mandating lifecycle water disclosure from hardware manufacturers represent an emerging but critical dimension of comprehensive water stewardship in AI infrastructure (Odhabi & Abi-Raad, 2024).

11. FUTURE OUTLOOK

One of the biggest trends in the future of data center growth is hyperscale AI centers. These are conversions, scaled up, to support a workload driven by massive computations, deep learning, real-time analytics and cloud-based AI services. Hyperscale data centers can house thousands of processors, generating vast amounts of heat. Absent enhancements of cooling technology efficiency, water usage of facilities like this could increase tremendously(Stacciarini & Gonçalves, 2025). Thus, future infrastructure for space cooling will have to be based on innovative cooling technology that ensures low freshwater consumption and high performance.

Another notable the trend is the rising utilization of AI based data center operation management. Machine learning algorithms can detect temperature, humidity and workload distribution instantly, enabling a more efficient running of the cooling system. Predictive control-user Cooling States such as Cooling Levels may be adjusted prior to overheating, reducing the waste of water and energy. By applying AI to its own infrastructure to achieve efficiency gains without cutting other computational capability, these benefits can be quantified. This transition in traditional fixed cooling systems towards adaptive systems enabling dynamic response to time-varying environmental conditions.

The future might also hold increased use of alternative cooling techniques with less reliance on freshwater use. With close-loop liquid cooling, immersion cooling, seawater cooling and other technologies expected to see increased adoption in the face of stricter environmental regulations and sustainability defined as the panacea for technology companies. In cold regions, natural air cooling can be used for much of the year, thus greatly saving both energy and water. Moreover, studies exploring novel materials and heat transfer methods with potential to further eliminate cooling needs are underway.

The rising adoption of renewable energy sources will also influence the future of data centers. Solar, wind, and hydroelectric power are already filling the U.S. energy grid needs with less carbon emissions, but they use water in a different way, than non-renewable plants do. Thermal generation of electricity requires lots of water for cooling, renewable energy sources such as wind, solar, hydro have little or no water requirements. As more data centers adopt renewable energy, their indirect water footprint that stems from the generation of the electricity they consume could diminish. Yet renewable energy systems themselves will have to be carefully designed so as not to cause new environmental problems.

And it is always expected that government regulations and international standards will weigh heavier in the future. Several nations have started acknowledging the negative effects on the environment due to digital infrastructure and are crafting policies to encourage sustainable design. These measures might cover water usage restrictions, efficiency reporting obligations, and credits for adoption of low-water cooling methods. Global norms could, in principle, be defined also by international agencies to guide the spread of artificial intelligence in a way that would avoid unnecessary depletion of natural resources.

However, the future of AI infrastructure does not have to bring with it a new wave of environmental destruction. Advanced cooling techniques, better efficiency models, and robust governance frameworks will enable the continued expansion of AI with an evolving water footprint. The secret lies in flexibly combining technology development with environmental planning (via environmental impact assessment) from the beginning of the infrastructure development process.

Figure 5. Shows the projected increase in AI infrastructure demand in the coming decades

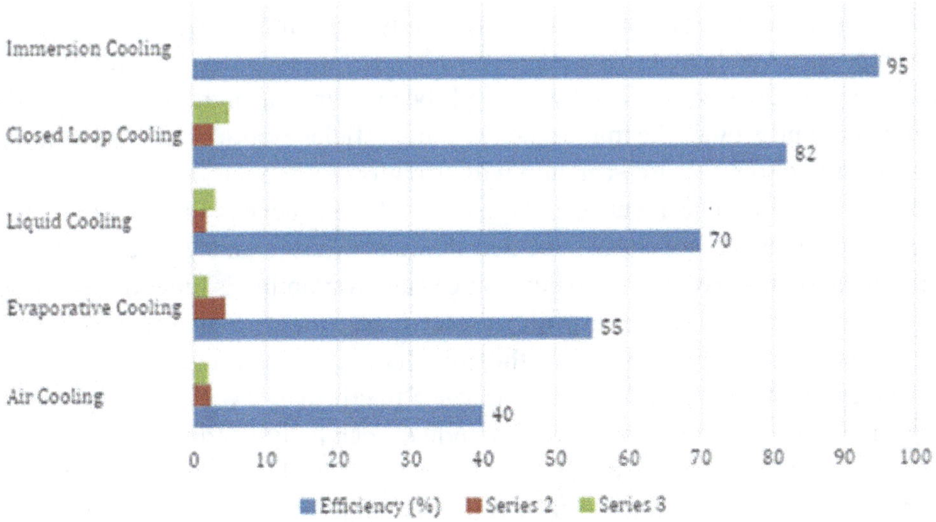

To sum up, the sustainability of the future of AI data centres will depend on the collective capacity of engineers, policy makers and the organisation to coalesce around sustainable solutions. As artificial intelligence continues to get more powerful and more prevalent, so too will the responsibility to mitigate its impact on the environment. Through efficient cooling systems, sound policies, as well as through incorporating environmental constraints in the planning stages – the international community should be able to realize the promises of A.I, without endangering the long-term availability of vital natural resources such as water.

AI computing demand is expected to increase dramatically over the next decade. As data center infrastructure expands, water management will become a critical sustainability challenge. Emerging technologies such as immersion cooling, AI-optimized thermal management, and seawater cooling systems offer promising solutions. However, technological innovation must be complemented by environmental governance and sustainable planning.

12. CONCLUSION

The explosive growth of artificial intelligence has reshaped the digital world and driven an unprecedented need for high-performance computing hardware. And at the heart of it all: the modern data center, which serves as the backbone for machine learning, cloud computing, and large-scale AI applications.

In this chapter, we analysed the expansion of AI data centers in relation to increasing pressure on fresh water resources globally. The findings revealed that water consumption is highly tied to the intensity of computation, weather conditions, and the cooling system design. As AI workloads become increasingly complex and the need for cooling grows, the importance of water efficiency as a determinant for the sustainability of digital infrastructure will also grow.

The worldwide placement of data centers also influences the environmental impact. Facilities are sometimes located in areas of the country where power is cheap or data connectivity is good but where water availability is limited. Therefore, proper site selection and environmental impact management is crucial to preventing unsustainable resource utilization in the growth of AI infrastructure.

Mathematical tools such as Water Usage Effectiveness (WUE), water–energy efficiency ratios and regional water risk indices, can be utilized to assess the environmental performance of data centers(Gao et al., 2021). These models provide researchers and policymakers with an understanding of how to compare various cooling technologies, project water use, and consider alternatives that reduce environmental consequences.

The pace of technological innovation in reducing data center energy use is promising. High-performance cooling solutions like liquid cooling, immersion cooling, closed-loop systems, and AI-based thermal management have the potential to decrease water use by orders of magnitude while still enabling high levels of computational performance. the employment of renewable energy sources and the application of enhanced building design can mitigate both the direct and indirect water use. While some of these technologies have higher upfront costs, they deliver long-term gains by increasing efficiency and enabling sustainable operation.

REFERENCES

Alsouda, F., Bennett, N. S., Saha, S. C., Salehi, F., & Islam, M. S. (2023). Vapor compression cycle: A state-of-the-art review on cycle improvements, water and other natural refrigerants. *Cleanroom Technology*, *5*(2), 584–608. DOI: 10.3390/cleantechnol5020030

Bashroush, R., Rteil, N., Kenny, R., & Wynne, A. (2020). Optimizing server refresh cycles: The case for circular economy with an aging moore's law. *IEEE Transactions on Sustainable Computing*, *7*(1), 189–200. DOI: 10.1109/TSUSC.2020.3035234

Brundage, M., Avin, S., Clark, J., Toner, H., Eckersley, P., Garfinkel, B., Dafoe, A., Scharre, P., Zeitzoff, T., & Filar, B. (2018). The malicious use of artificial intelligence: Forecasting, prevention, and mitigation. *arXiv preprint arXiv:1802.07228.*

Clarke, T. (2025). Sustainability in international public policy and governance.

da Silva, L. A., Vasconcelos, E. S., de Paiva, L. F. R., Collaço, M. H. d. V. R., Joaquim, W. M., de Lima, A. D., Goulart, C. S., & Teixeira, E. P. (2025). ENERGY CONSUMPTION OF DATA CENTERS AND THE PRIMARY ENERGY METRIC: GLOBAL ANALYSIS 2000–2024 AND PROJECTIONS UNTIL 2030. *Editora Impacto Científico*, 21-41.

du Plessis, A. (2023). Water resources from a global perspective. In *South Africa's Water predicament: Freshwater's unceasing decline* (pp. 1–25). Springer. DOI: 10.1007/978-3-031-24019-5_1

Gao, J., Shao, C., Chen, S., & Zhang, X. (2021). Spatiotemporal evolution of sustainable development of China's provinces: A modelling approach. *Ecosystem Health and Sustainability*, *7*(1), 1965034. DOI: 10.1080/20964129.2021.1965034

Gehring, J. C., & Cand, B. (2024). *The Sustainable Energy Imperative: A Future Generations Perspective on Technologies Leading the Clean Energy Transition*. The Global Youth Council on Science, Law & Sustainability.

Haghdadi, N., Cizek, P., Hodgson, P., Tari, V., Rohrer, G., & Beladi, H. (2018). Effect of ferrite-to-austenite phase transformation path on the interface crystallographic character distributions in a duplex stainless steel. *Acta Materialia*, *145*, 196–209. DOI: 10.1016/j.actamat.2017.11.057

Masanet, E., & Lei, N. (2020). *How much energy do data centers really use*. Aspen Global Change Institute.

Medarac, H., Magagna, D., & Hidalgo Gonzalez, I. (2018). Projected fresh water use from the European energy sector. *Disaggregated Fresh Water Withdrawal and Consumption in the EU up to, 2050.*

Mytton, D. (2021). Data centre water consumption. *npj. NPJ Clean Water*, *4*(1), 11. DOI: 10.1038/s41545-021-00101-w

Organization, W. H., Programme, U. U. N. E., & Health, W. O. f. A. (2022). *One health joint plan of action (2022–2026): working together for the health of humans, animals, plants and the environment.* World Health Organization.

Silander, D. (2019). The European Commission and Europe 2020: Smart, sustainable and inclusive growth. In *Smart, sustainable and inclusive growth* (pp. 2-35). Edward Elgar Publishing.

Stacciarini, J. H. S., & Gonçalves, R. J. A. F. (2025). Data Centers, Critical Minerals, Energy, and Geopolitics: The Foundations of Artificial Intelligence. *Sociedade & Natureza*, *37*(1), e77215. DOI: 10.14393/SN-v37-2025-77215

Vaughn, M. (2021). ASHRAE Research Report: 2020-2021. *ASHRAE Journal*, *63*(10), 73–87.

Zhang, G., Zhang, S., Zhang, W., Shen, Z., & Wang, L. (2020). Distributed energy management for multiple data centers with renewable resources and energy storages. *IEEE Transactions on Cloud Computing*, *10*(4), 2469–2480. DOI: 10.1109/TCC.2020.3031881

Chapter 6
Legal Battles Over Water for AI:
Landmark Cases and Future Precedents

Naveen Nandal
http://orcid.org/0009-0007-2272-7437
Delhi Institute of Higher Education, India

Reema Bhattacharya
http://orcid.org/0000-0001-7687-2479
Jagannath University, India

Arti Sharma
http://orcid.org/0009-0006-2074-8157
Jagannath University, India

Suman Paliwal
Jagannath University, India

Dhanvanti Godha
Jagannath University, India

ABSTRACT

The swift growth of artificial intelligence (AI) infrastructure has introduced important but less explored challenges of water governance. AI water use is growing rapidly and causing an additional strain on other regions & populations that place on water supplies. Through landmark controversies on water use by AI, this chapter examines the global legal controversies on water use by AI and the evolving regulatory responses. The study applies a qualitative methodology based on case law analysis

DOI: 10.4018/979-8-2600-2761-5.ch006

and legal documents to look at how existing legal doctrines like public trust doctrine, beneficial use, precautionary principle, and the right to water are being adapted to such challenges. The focus on environmental justice, corporate power and weak regulation was highlighted. This course will further analyse the way courts and policymakers are creating new legal precedents either to protect technology or the environment. The conclusion of this chapter sets out key legal and policy pathways for responsible and equitable water governance in the AI age.

INTRODUCTION AND BACKGROUND

In recent years, artificial intelligence (AI) has transformed from a specialized technology to a fundamental infrastructure of the global economy, politics, and society. This change has come together with an explosive expansion in data centers and high-performance computing facilities to support machine learning, cloud services, and large-scale data processing. Despite its growing environmental and legal importance, the water use of infrastructure of this sort has attracted surprisingly little attention, despite the focus of research and policy debate on its energy use. To keep high-performance machines and computer hardware from overheating, AI data centers additionally rely on cooling systems that consume water. Millions of liters of freshwater are used annually for these processes (Jones, 2018; Van der Veen & Smith, 2022), often in water-stressed or ecologically vulnerable regions. As people learn more about the water footprint of digital systems, the understanding of AI continues to shift away from one of "virtual" or "immaterial" technology, and towards one with a footprint. People living near major data centers are troubled by their groundwater depletion, competition with agricultural and domestic water uses, and the sustainability of data centers. Often, these worries have led to lawsuits involving local governments, environmentalists, regulators, and global tech firms. According to Masanet et al. (2020) and Crawford (2021), the various conflicts indicate that the issue at stake is not merely the allocation of resources but rather a conflict to maintain the technology at the cost of nature or the communities.

These new disputes under the law have become more complex because of existing water governance. No laws have been made for digital infrastructure. The current regulations governing water distribution favor agricultural, industrial, and household purposes more than the environment. With the entry of AI infrastructure, such classifications get challenged due to the kind of industrial activity that yields an intangible output but has a high demand for resources, while also being concentrated in a geographical area. The regulating bodies and courts are asking a new question: Should AI data centers be considered essential infrastructure like a public utility, or should they be treated as commercial businesses and be bound by DRP norms?

How do we define fair apportioning of water in a world of digital development and human beings? Lawyers can help legal systems manage the conflict between sustainability and economic development as technology is developing ever so quickly. These issues have become even more complex with the development of artificial intelligence. In order to optimize performance, regulations, and resources, data centers are located in varied places. Water-related conflicts are governed and resolved in sharply different ways under transnational distribution. In sincere contexts, where legal challenges are raised, stricter permitting or transparency requirements create cancellation or greater challenges. In jurisdictions where regulations are weaker, it is more difficult for communities to challenge water allocations and secure protections for the environment. The inequality causes serious concerns over environmental justice, regulatory arbitrage, and global dumping of AI infrastructure-related impact. (Borgesius et al., 2022; Li et al., 2023).

There have been many known cases in the recent past that make this clear. The legal challenge to the allocation of permits for the withdrawal of water for data centers is underway, along with corporate water use data disclosures intervention, and efforts for weak ecosystems and vulnerable communities. Cases of this nature will set future precedents for the government regulating activities involving the consumption of water by AIs. Law is always a good way of looking at tensions among competing objectives, such as innovation, development, environmental protection, and equitable distribution. Moreover, the battles in court are connected to larger supply-chain and environmental problems. Due to climate change, an ever-growing population, and diminishing water resources, pressure on freshwater resources is increasingly rising. The problem of the environmental footprint of AI infrastructure, in particular, its growing dependence on freshwater resources, is also supported by the agreements with the literature in the area of Information and Technology. Recent evidence has shown that hyperscale AI data centers consume large quantities of water to cool systems, power generation, and operational maintenance, especially in regions where water shortage and climate stress have already been experienced. Xiao et al. (2025) estimated that worldwide AI-related water withdrawal could grow tremendously by 2030 as a result of the swift growth of the generative AI systems and high-performance computing facilities. Similarly, Mytton (2021) argued that the digital infrastructure is emerging as a factor of concern regarding global water insecurity, but it is understudied. Research by Li et al. (2023) also hints at the fact that AI-based cloud computing has more pressure on the local ecosystem because of the high number of hyperscale data centers that are located in areas with a low level of environmental protection. Another study conducted by the International Energy Agency (2024) also highlights the fact that water consumption by data centers will keep on rising along with the rising electricity demand due to artificial intelligence applications. Another point that Crawford (2021) makes is that the material and ecological aspects of AI systems are usually obscured by the narratives about digital technologies as clean or immaterial. Also, Van der Veen and Smith (2022) have indicated that the absence of transparency when it comes to corporate water consumption limits effective environmental stewardship and social responsibility. Another caution issued by UNESCO (2023) is that as competition intensifies over the freshwater resources, it is likely that legal tussles will escalate between industrial development and environmental protection, and community water rights. Together, these studies demonstrate that AI infrastructure is not just a technological phenomenon but a critical environmental governance challenge that requires the reinforcement of legal regulation, policy frameworks that are focused on sustainability, and corporate responsibility.

Figure 1. AI Infrastructure–Water Governance Nexus

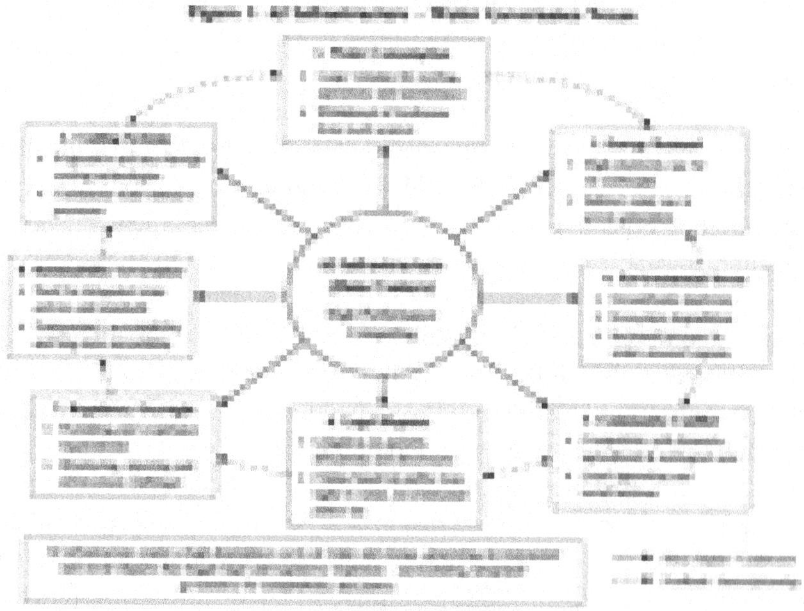

Considering the above-mentioned situation, the representatives of the Surveillance Incidence (SI) department on the ground have made some comments concerning the foundation of a mobile phone manufacturing factory in the area. The legal systems should do much more than settle disputes. Another area in which they can assist is developing smart regulatory frameworks to make AI avert systemic risk. The Chapter comes into the field through the globalization of the legal controversies on water use to support Artificial Intelligence (AI) infrastructure. The outlay of a ratified system attempts to evaluate landmark cases to learn about judgment reasoning and pattern controlling reaction among concurrentness across jurisdictions. The object of this Chapter, then, is to show how law either accommodates or fails to accommodate the life of interactive communication. Furthermore, the Chapter illustrates that legal problems involving AI in water matters are not merely technological, regulatory, or justice problems but instead about power, governance, and justice. Decisions in water allocation embody more of the larger social priorities and values - they reflect on who makes decisions on behalf, who is willing to listen, and how risks and benefits are distributed. Consequently, these conflicts are indicative of the evolving conditions of technology, law, and environmental sustainability. In conclusion, the rapid development of AI infrastructure has put water governance on the front burner of the discussion. With all of this, and the courts, regulators, and communities wrestle with the ramifications of all this, a new set of legal precedents is being defined to ensure sustainable digital development in the future. I will discuss this in this Chapter, starting with the conceptual and legal framework of water governance within AI.

AI DATA CENTERS AND THE EMERGING WATER GOVERNANCE PROBLEM

The rapid implementation of the artificial intelligence (AI) technology has been accompanied by a rapid physical growth to support the technologies, in particular, hyperscale data centers. The core of the AI systems most frequently utilized nowadays, due to their ability to achieve massive computing, cloud storage, and real-time processing, are these facilities, which often cover large areas and accommodate thousands of high-performance machines. The online products of these systems appear to be intangible. In actuality, their mode of operation involves the utilization of the physical resources, which in this case are energy and water. The most significant yet least controlled aspect of AI infrastructure is water consumption, which is associated with complex governance challenges on the local, national, and global levels (Masanet et al., 2020; Li et al., 2023). Water also assists AI data centers to operate well, as well as cooling their hot-generating computing devices, which are tightly packed together to avoid malfunction. Increasingly, such relatively water-consuming yet efficient water cooling technologies are substituting or supplementing traditional air-cooling technologies. These systems are often reliant on the continuous movement of the freshwater in the municipal supply, groundwater sources, or other water bodies. The larger of these facilities uses millions of liters of water daily, particularly when peak computational loads occur. The increasing energy demand in the world is further straining water. This is particularly evident in locations where there is already a shortage of water, where a drought occurs, or the water demands of the agricultural sector and household consumers are competing (IEA, 2022; UNESCO, 2023).

The increasing pressure to do something about climate change, and regulatory systems that had never been designed to take into consideration the environmental effects of digital infrastructure, is a problem in governance. Current water management frameworks view water users through the traditional lens of agriculture, industry, and domestic water supplies. These categories, however, were not meant for AI data centers. Depending on the regime, whether they are classified as industrial plants or as critical infrastructure varies. The classification determines the issue, monitoring, and enforcement of water permits. The lack of clarity creates regulatory loopholes that the stakeholders will exploit or challenge, resulting in disputes among them (Borgesius et al., 2022). Controlling the water use of AI infrastructure is complicated by the lack of information about how much water major corporations use. Business regulation and water companies' disclosures of operational data have confirmed that technology companies do keep detailed data on their use. However, this lack of transparency is not in line with growing community, environmental, and regulatory demands for accountability and the public disclosure of their plans. Most times, a lack of reliable data hinders efficient environmental impact assessments and the stakeholder capacity to evaluate the sustainability of proposed or existing data center projects. According to Crawford (2021) and Van der Veen and Smith (2022), there is increasing litigation around this topic, wherein litigants are forcing disclosure or challenging the adequacy of environmental disclosure.

AI data centers are sometimes situated in regions that provide economic incentives or favorable regulatory conditions. Others are established in places with access to abundant resources, such as water, or in systems with a low risk of natural disasters. Contrary to one's assumptions, some of these places are actually water-stressed regions where the installation of water-intensive infrastructure increases vulnerability. Local communities are now crying foul that water that is set aside to serve data center activities is taking away their own share of the water to serve the basic needs. The controversies lead to an important question: whether the water consumption of digital innovation and economic development should take priority over human, agricultural, and ecosystem needs or vice versa? Courts and regulators usually resolve these conflicts with no clear legal duties (UNESCO, 2023). The issue is far more serious because the development of AI has been globalized. International technology companies have the largest residual legal system in the world that operates across various conflicting competing states that have competing laws. This presents opportunities for 'regulatory arbitrage' whereby companies can site data centers in jurisdictions with less stringent environmental regulations or weaker enforcement capacity. Although it results in operational savings, these strategies could be used to impose environmental costs on the communities that are not prepared to resist or manage the initiative. The discussions of access and uneven distribution of environmental risk produced by infrastructures will inevitably be oriented by AI governance (Li et al., 2023).

Climate change also adds complexity to the governance of water to support AI. Due to rising temperatures and unpredictable weather patterns, there is an anticipation of increased water scarcity in a number of countries. The warmer climate increases cooling demand in data centers, thus leading to an increase in water consumption. This creates a feedback loop, which also contributes to and suffers from environmental harm, the underlying AI overlay infrastructure. According to the IEA (2022), future uncertainties will also pose a challenge to policymakers and regulators. The governance issues that may arise as a result of the use of AI in the water sector are not only technical or administrative in nature but also are entangled in more general legal, ethical, and public policy concerns. The choices made in how to allocate water show the importance of different values for society, which may also prioritize between economic development and environmental sustainability, as well as upholding vulnerable groups, and whether water is to be considered a human right. There may be interesting legal conflicts over the use of water by AI, as sites where these contending values are asserted and negotiated. The paper discusses how societies are starting to respond to the environmental effects of a digital transformation and the possible usefulness of applicable current laws.

With these new emerging concerns in mind, even some jurisdictions are starting to explore new regulatory strategies to improve the regulation of AI water use. They have put in place regulatory measures, which will involve more stringent conditions regarding the environmental impact assessment, water permits, and incentives for the adoption of water-efficient technologies. However, these would not always prove effective, and in most cases would be inappropriate to the problem concerned. As a result, courts continue to play an important role in governance by providing a forum through which disputes may be resolved, and so important precedents are set. The challenge of water governance concerning the AI data centers is not just technical. It also involves legal, environmental, and social factors. Solutions are required that incorporate pertinent laws, policy alternatives, and regulatory changes. Recently, the development under WPR has given rise to an ongoing debate on the identification of the various legal regimes governing the space.

Table 1. Key Governance Challenges in AI Water Regulation

Governance Challenge	Legal Concern	Environmental Impact	Regulatory Gap	Suggested Reform
Excessive water consumption	Allocation disputes	Groundwater depletion	Weak disclosure rules	Mandatory reporting
Lack of transparency	Trade secrecy claims	Poor accountability	No uniform standards	Public disclosure frameworks
Regulatory ambiguity	Classification disputes	Inconsistent oversight	AI infrastructure not legally defined	Specialized regulation
Environmental injustice	Unequal burden on communities	Water insecurity	Weak community participation	Participatory governance
Climate vulnerability	Drought-related conflicts	Ecological stress	Static permit systems	Dynamic water permits
Corporate concentration	Resource monopolization	Local inequity	Weak enforcement	ESG-linked compliance

SECTION 3: LEGAL AND CONCEPTUAL FRAMEWORKS

Legal and Conceptual Frameworks

To comprehend the litigation associated with the use of water in artificial intelligence infrastructures, this project draws on a synthesized analytical framework based on environmental law, technology regulation, human rights, and administrative law. The concurrence of these areas indicates that AI dispute waters are not just technical disputes over resources. Instead, it is a legal struggle that is based on the dissimilarity of the normative frameworks, institutional capacity, as well as the variance and evolution of the public interests. This section describes the critical conceptual and regulatory frameworks of governance in relation to water to support AI data centers, which will support the analysis of landmark litigations and set trends. The majority of the disagreements regarding environmental law are over water rights and the water allocation doctrine. In different jurisdictions, water law is founded on various principles, which include, but are not limited to, prior appropriation, riparian rights, and state ownership. In most regions, water has been treated as a common good, and it is subject to the regulation of the state. Moreover, the decision to allocate is based on the need, sustainability, and equity. However, AI infrastructure confronts these traditional structures by establishing a new category of non-population high-volume water users whose value to society is still disputed. Increasingly, courts are being requested to interpret whether the use of water to support digital infrastructure qualifies as beneficial use or is subject to public interest, both of which started their development in the context of agriculture and industry (Getches, 2017; Dellapenna & Gupta, 2020).

The Public Trust doctrine is also linked to environmental law. Under this doctrine, certain natural resources, such as water, are preserved for public use. The government, meanwhile, must protect these assets for present and future generations. This doctrine was invoked in a number of environmental cases in which the State's decision prioritized private or commercial over ecological sustainability and public access. The public trust doctrine can be a significant legal tool when it comes to contesting whether vast withdrawals of water at AI data centers agree with the state's obligation to protect water and other natural resources. In regions affected by climate-change-induced water scarcity, there will also be calls to consider concepts of intergenerational equity (Sax, 1970; Wood, 2014). The law of technology and infrastructure, at the same time, determines the manner in which permission is received by the data centers of AI, they are regulated, and they are monitored. Toxic plant laws often limit where plants can be built, often taking environmental impact assessments and industrial permits into account as well. Regulatory systems trail far behind the advancements in digital technology. At times, data centers that boast a large carbon footprint from AI systems are subject to a general industrial or general commercial status that does not take their unique resource use into account. The effect of this mismatch will be gaps that result in inconsistent enforcement, legal uncertainty, and greater recourse to litigation to resolve disputes (Borgesius et al., 2022).

Administrative and regulatory law includes the legal framework that governs public authorities that grant permits, enforce environmental standards, and settle water disputes. Water allocation disputes before administrative agencies happen due to procedural flaws, failure to take into account environmental factors, and violation of statutory mandates. The legal analysis is based on the issues of public consultation, transparency, and the issues associated with accountability. Water has been identified in human rights as a fundamental human right, and the water-related burdens are already concentrated on already burdened populations in the environmental justice framework. The United Nations considers the right to safe water a human right that establishes the normative standard of legal and policy. The unequal distribution of costs and benefits of AI development has attracted the attention of environmental justice perspectives, whereby the disadvantaged communities are usually the ones who bear the environmental costs, yet receive few economic benefits (UN General Assembly, 2010; Schlosberg, 2013). The decision-making process of NCAP takes into consideration all aspects of the environment, economic, and social. It emphasizes the need to manage resources in the long term, taking precautionary steps, and internalizing environmental costs. Conflicts around data centers can be seen as a conflict between individual interests and the general will to achieve sustainability objectives. The responsibility of technology companies to reveal their impact on the environment and run a socially responsible business is corporate accountability and transparency. Conflicts in law occur when corporations refuse to release detailed information related to their water use, citing trade secrets. Public access to information is essential for informed decision-making. Many times the same theme comes in legal fights, i.e., between business secrecy and public responsibility when it comes to AI water use (Crawford, 2021).

AI technology is developing on a global scale, and thus, working with comparative law is important. In addition, environmental regulation and water governance vary significantly from jurisdiction to jurisdiction. Recognizing more harmonized governance approaches requires a comparative analysis of existing styles, patterns, and best practices. Despite this, inconsistent regulatory standards risk undermining collective efforts towards regulating the environment, as per Dellapenna & Gupta (2020). Multiple domains of law and governance are involved in the legal battles over AI water. Environmental law allows for the allocation and protection of resources, then technology and administrative law will help with infrastructure development-related matters, plus human rights and environmental justice will bring in equity and fairness, and through sustainability governance will bring about the ability to balance competing interests, future-wise. This Chapter explores cases relevant to landmark litigation and future litigation; this will facilitate legal innovation and reform necessary to solve AI-based consumption of water.

Figure 2. Legal Framework Governing AI Water Use

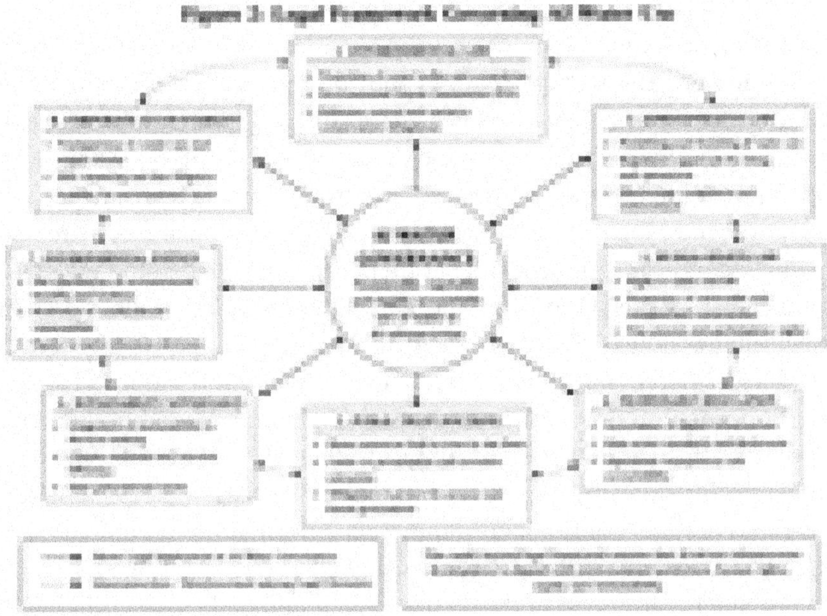

Table 2. Comparative Legal Principles Applied to AI Water Governance

Legal Principle	Jurisdictional Examples	Application to AI Infrastructure	Emerging Judicial Trend
Public Trust Doctrine	USA, India	Limits privatization of water resources	Greater ecological oversight
Precautionary Principle	EU, India	Restricts uncertain environmental risks	Preventive governance
Beneficial Use Doctrine	USA	Questions legitimacy of high-volume AI water use	Sustainability-based reinterpretation
Right to Water	South Africa, Chile, India	Protects community access to water	Rights-based environmental review
Environmental Justice	USA, South Africa	Examines unequal burdens on vulnerable groups	Equity-centered governance
Corporate Transparency	USA, EU	Requires disclosure of water consumption	Expanding accountability norms
Sustainable Development	Global frameworks	Balances innovation with ecology	Integrated governance approach

Landmark Legal Battles: Doctrinal and Comparative Analysis

The legal conflicts concerning water use for AI infrastructure become most analytically illuminating through case studies and court judgments. Lawsuits centered on the use of water for artificial intelligence are still developing; however, existing lawsuits touching on environmental law and water law already offer a strong doctrinal framework for their interpretation. Disputes over water use arising from the use of AI are not an entirely new legal issue. They are rather a slow extension or re-realization of an existing legal norm. At the moment, there is a battle that could make a lot of folks a whole heap smarter. Google's data center activity in The Dalles, Oregon, has led to lawsuits under state public records laws to compel the release of the company's water use. The deliberation stemmed from whether that information was protected commercial information or whether it was a disclosure in the public interest. The legal rationale in this context sounds similar to that of the US Supreme Court in Food Marketing Institute v. Argus Leader Media, 588 U.S. _(2019), wherein the court construed the term 'confidential commercial information' generously. The fact that Google has revealed some of its water usage data is interesting and represents an opposite trend already seen in the courts, where companies are utilizing public resources (especially essential ones like water), and confidentiality takes a back seat to transparency and accountability. As a result, this case shows an emerging doctrinal shift in which transparency about the environment is increasingly regarded as a condition of corporate legitimacy.

The doctrine of beneficial use, which is a key element of prior appropriation systems, relates to whether the water use by the data center in the American West is beneficial. In *Arizona v. California*, 373 U.S. 546 (1963), the first-ever case that mandated water allocation for equitable and beneficial use. Recent arguments in Arizona about plans for hyperscale data centers in a water-stressed area beg the question: Does AI infrastructure satisfy this doctrinal requirement? As of now, courts have not definitively ruled on the matter regarding AI. However, the obstructions and public opposition are becoming ever more pronounced, with it being clear that economic productivity should no longer suffice as justification for the extraction of water, at least not if it harms nature or humanity's need for water (UNESCO, 2023). This refers to a progression in the beneficial-use standard from economic to environmental and social beneficial use. The public trust doctrine has evolved to support this trend, National Audubon Society v. Superior Court, 33 Cal. 3d 419 (1983). In the Mono Lake case, a California Supreme Court decision 3d 419 (1983), the court ruled that the state has an ongoing obligation to monitor water allocations in order to protect the public trust resources. This principle implies that even the previously granted water rights can be reconsidered in case they impact the ecology. When it comes to AI data centers, the doctrine implied that the breach of municipal water law will occur when the governments ever assign large amounts of water on a permanent basis to a private technological company in a way that is incompatible with the public law. The government cannot lose its mandate to control the management of natural resources towards the common good. This can be seen in Illinois Central Railroad Co. v. Illinois, 146 U.S. 387 (1892). In effect, the amalgamation of these cases offers a resounding doctrinaire support of re-evaluation of long-term water allocations in the light of impending environmental pressure emerging out of AI infrastructure.

An alternative perspective, which is the precautionary principle, is provided by the European law. In a case of uncertain impact, Waddenvereniging and Vogelbeschermingsvereniging v. Staatssecretaris van Landbouw, Case C-127/02 (2004), the Court of Justice of the European Union held that projects of uncertain effect must be carefully reviewed and may be limited where risks are not exclusionable. Such jurisdictions as the Netherlands have already started to impose limits on hyperscale data centers due to concerns about the potential cumulative effect. On the same note, in Case C-461/13, Bund fur Umwelt und Naturschutz Deutschland (Weser River Case) (2015), it was established that there should be strict control over any water quality deterioration under the EU law. The above-mentioned cases suggest a new environmental governance that is preventive, meaning the absence of certainty does not justify inaction, but caution. In places where constitutional environmental rights are well developed, courts have used a more rights-based approach. The Supreme Court of Chile has, in Case Rol No. 5888-2019, acknowledged access to water as part of the constitutionally protected rights to life and a wholesome environment. Even though it is not about AI infrastructure directly, this case law sets the ground for future litigation where industrial water use, including that of data centers, could be challenged on grounds of excessive infringement of fundamental rights. The process of producing water should be proportionate to the basic human requirement for water.

The Indian legal system gives perhaps one of the widest interpretations to Environmental Rights. In Subhash Kumar v. State of Bihar (1991) 1 SCC 598, the Supreme Court held that the right to life under Article 21 includes the right to clean drinking water. The Supreme Court further confirmed this principle in *M.C. Mehta v. Kamal Nath* (1997) 1 SCC 388, where the Court applied the public trust doctrine to restrict a private person's exploitation of natural resources. In addition, Vellore Citizens Welfare Forum v. Union of India (1996) 5 SCC 647 introduced the precautionary principle in our national law, which states that harm must be prevented even if there is no complete scientific evidence. This case law builds a legal framework for judicial review of AI-related water usage that harms the public or the environment. Another dimension of real environmental enforcement is the case of Friends of the Earth v. Laidlaw Environmental Services, 528 U.S. 167 (2000), which was decided by the Supreme Court of the United States, which recognized the roles of citizen litigation. The Court held that the risk of environmental harm and violations of law is enough to confer standing. Its implication is significant for water governance in AI. The ability of civil society and community members to hold tech companies accountable is becoming real. Furthermore, such practices may take place even in cases where direct harm is hard to gauge.

The Constitutional Court of South Africa eventually ruled on Mazibuko v. City of Johannesburg (2009) ZACC 28 in a socio-economic right context on the right to water. Although the Court adopted a deferential stance on the state policy, it nevertheless declared that the right to water is justiciable. The issue here is one of equity and distributive justice among the water governing institutions and decision-makers. Especially when the vulnerable communities disproportionately feel the benefits or impacts of any decision. Overall, the legal governance of water for AI is not taking place in a vacuum. Rather, it is developing as a reinterpretation of existing legal doctrines across both common and civil law jurisdictions. Such values as beneficial use, public trust, precautionary governance, environmental rights, and transparency are being adjusted to the particular challenges of digital infrastructure, sometimes in a gradual way and sometimes more coercively. The courts' alleged politics are often blamed for the system's failures of strident social justice. Nevertheless, the above actually indicates that it is a more intricate affair. This demonstrates that the legal environment is heterogeneous and dynamic. There are also some jurisdictions that rely more on constitutional rights or procedures, and there are those that are more based on prevention and sustainability. However, the two seem to be moving in a similar direction; i.e., incorporating the environmental concerns into the management of AI infrastructures. With the lawsuits continuing to increase, case law in its initial days is likely to take the form of an ever-evolving corpus that will lead to responsible and sustainable development of AI. Table 1 offers a comparative legal analysis of AI water governance cases across jurisdictions, which demonstrates how various legal principles, including public trust, precautionary regulation, and beneficial use, are applied to judicial reasoning.

Table 3. Comparative Legal Analysis of AI Water Governance Cases

Jurisdiction	Case	Legal Principle	Relevance	Key Impact
USA	Food Marketing Institute v. Argus Leader Media (2019)	Transparency vs trade secrecy	Disclosure of water use	Expands accountability
USA	Arizona v. California (1963).	Beneficial use	Water allocation	Sustainability reinterpretation
USA	National Audubon Society v. Superior Court (1983)	Public trust	Water rights review	Protects ecosystems
USA	Illinois Central Railroad v. Illinois (1892).	Public trust	Resource ownership	Limits privatization
USA	Friends of the Earth v. Laidlaw (2000)	Enforcement	Citizen litigation	Strengthens compliance
EU	Waddenzee Case (C-127/02)	Precautionary principle	Risk regulation	Preventive governance
EU	Weser River Case (C-461/13)	Water protection	Resource degradation	Strict limits
Chile	Roll No. 5888-2019	Right to water	Industrial vs public use	Rights-based approach
India	Subhash Kumar v. State of Bihar (1991)	Right to water	AI disputes	Expands Article 21
India	M.C. Mehta v. Kamal Nath (1997).	Public trust	Resource protection	Restricts exploitation
India	Vellore Citizens Welfare Forum (1996).	Precautionary principle	Environmental risk	Sustainability focus
South Africa	Mazibuko v. Johannesburg (2009).	Right to water	Equity	Social justice

The recent real-life scandals of overly large data centers only further indicate how much tension is building up between artificial intelligence infrastructure and sustainable water governance. In the Dalles, Oregon, citizens and environmental activists questioned the secrecy of Google's water consumption after it was revealed that the company's data centers used millions of gallons of municipal water each year. This was followed by litigation battles and public records battles on whether the information about the water use of corporations should be kept confidential or released in the public interest. The scandal also brought up questions of transparency, social accountability, and allocation of scarce water resources during droughts in drought-prone regions (Crawford, 2021).

On the same note, local communities and environmental groups in West Des Moines, Iowa, have cast doubts on the sustainability of cloud and AI infrastructure, which is consuming a lot of water. The increase in the number of local water demands was reportedly caused by the expansion of data centers (IEA, 2024).

In the Netherlands, the Dutch government has provided restrictions on the development of hyperscale data center projects, including those related to Meta Platforms, due to concerns over excessive energy and water consumption, land use pressures, and environmental sustainability. The aversion of the population and regulation were indicative of a precautionary tendency of governance whereby the risks to the environment were deemed to be sufficient factors to restrict further development, even in situations when the economic benefit was high (Borgesius et al., 2022).

It has also been growingly controversial regarding the development of the data centers in Santiago and the surrounding regions that are characterized by years of drought conditions. Further certification of the digital infrastructure with high water consumption, according to the claims of the civil society groups, threatens ecological sustainability and local water security. These conflicts are already becoming mixed between environmental protection and constitutional rights to water and to a healthy environment (UNESCO, 2023).

Collectively, these examples allow us to conclude that the conflicts over the AI infrastructure are no longer hypothetical or purely doctrinal. Rather, they consider the new conflicts of global governance whereby, in such disputes, courts, regulators, corporations, and communities are arguing out the new relationship between digital innovation, environmental sustainability, and resource justice.

Key Themes and Judicial Insights

Water uses for AI systems are becoming the subject of litigation. These cases show recurring themes that show how legal systems are evolving concerning the impacts of digitization. Themes include transparency and institutional adequacy, prioritization, and equity. A major theme is the redefinition of 'public interest' in water allocations. Historically, water law has prioritized essential uses, e.g., agriculture and domestic water use, by explicitly prohibiting or limiting attempts to divert or affect allocations for essential uses. However, AI infrastructure poses a challenge to essential uses prioritization because it produces diffuse, indirect, and spatially global economic activity at localized environmental cost. Courts are now deciding if data centers and AI water use are excessive or justified. Judicial reasoning tends to balance economic development with the environment (Dellapenna & Gupta, 2020). Regulatory decisions must be procedurally rigorous and environmentally sound, as courts are actively scrutinizing them. These courts are looking into whether EIA, public participation, water data transparency, and other submissions are true. In governance processes, rather than outcomes, are more significant to legitimacy according to an emerging shift in environmental governance. The broadening trend in environmental governance is a trend in environmental governance (Mashaw, 2018). Corporate secrecy and environmental transparency are in terrible conflict with each other. Firms developing apps and systems must not reveal their operations' data to the public. However, Courts and regulators are querying this stance in major pollution matters. Judicial reasoning suggests that there is a growing recognition of the need for greater transparency in cases where corporate sector activities affect public resources. It does not seem likely to be solved easily, but firm early rulings suggest a broad opening up of information with regard to the environment (Crawford, 2021; Borgesius et al., 2022).

As water scarcity increasingly affects certain groups, environmental justice and the fair distribution of resources are becoming more common in the courts and regulators. The presence of environmental burdens is not equal for everyone, as per the courts. Environmental justice is added to the reasoning of environmental law, which is making a trend these days, a modification and change in the environmental laws wherein now equity is accepted as an essential element in the sustainable governance (Schlosberg, 2013). Too often, the regulatory frameworks around the AI infrastructure are inadequate and insufficient. They came from an era when it was easy to classify industrial processes, and their physical impact was predictable. Since these data centers operate at the intersection of multiple sectors, this results in regulatory ambiguity and gaps, thus requiring legal reform. This judicial acknowledgment of the inadequacy of regulation is essential for effective case-by-case adjudication. Precautionary thinking has emerged in new contexts of scientific uncertainty, like large-scale water extraction for artificial intelligence infrastructure. In Water Scarcity Areas, Court Preference Goes for Environment over Economy. This change in environmental law addresses risk avoidance and sustainability for the future in the absence of certainty (UNESCO, 2023). There is new insight in judicial thinking regarding the link between environment and technology, and it requires a comprehensive assessment of the environment. According to Geissdoerfer et al., (2017), Courts found that decisions on AI infrastructure need to take into account the wider environment and societal context, which interrogates the traditional practice of assessing a project on a project basis. Due to the rapid growth of technology, courts serve as a de facto policymaker when the legislation does not provide appropriate guidance. The executive is already playing a significant role in policymaking by interpreting existing laws. This has resulted in immediate solutions to critical public issues. However, it raises important concerns with regard to the legitimacy of democracy in India and the balance of power. In water governance involving AI, the courts are key places for defining rules.

The legal field is undergoing a transition as courts face new challenges in regulating the environmental impact of artificial intelligence. In the earlier cases, interpretative frameworks and normative benchmarks were established, making the body of law more coherent. The precedents shape the expectations of policymakers, regulators, and industry actors the world over. Disputes between the public interest, integrity of the legislative/procedural processes, transparency, fairness/equity, and regulatory fitness cannot be narrowly defined. The analysis will shift to environmental justice and the rights of communities subsequently.

Environmental Justice and Community Rights

Disputes over water use in AI infrastructure are more than resource allocation disputes; they reflect environmental justice and community rights. AI data centers are growing rapidly around the globe, and any negative impact, including water consumption, is transferred to vulnerable communities. The uneven burdening of the environment counters institutional arrangements that favor economic outcomes over social and environmental ones. Environmental justice is compromised when big data centers are set up in water-scarce regions or those that are infrastructurally limited due to cost efficiencies, availability of land, or flexibility in rules. Environmental degradation follows due to competition over water resources, with conflict over resource availability with long-term impact (Schlosberg, 2013; Li et al., 2023). Statements of fairness and equity are common in disputes regarding access to water. Communities and advocacy groups say that AI-related infrastructure is being prioritized over water, disrupting the basic right to water and a healthy environment. Various international systems for the safeguarding of human rights recognize the right to safe water. Regulatory agencies and courts have to be cognizant of normative principles that are more generic, (United Nations General Assembly, 2010). The AI data centers raise matters of community engagement and procedural justice. People are worried about low consultation, transparency, and influence. Legal actions concern the duty of regulatory agencies to consult the public and hear them out. More courts are rejecting the legitimacy of environmental decisions that have procedural deficiencies. According to Mashaw (2018), distributive justice points to the unequal distribution of the environmental costs and benefits of the AI infrastructure, where costs are experienced locally, including withdrawal of water and stress on ecosystems. This creates a situation where communities incur development costs while benefiting little. Legal arguments on compensation, mitigation measures, or alternative resource allocation strategies for these imbalances vary in terms of the extent to which measures are mandated (Schlosberg, 2013).

AI water projects have faced opposition from Indigenous groups in certain areas who argue that these initiatives jeopardize critical water sources and infringe upon their rights. Familiarity with relevant legal protections is essential, such as domestic legal protections and the UN Declaration on the Rights of Indigenous Peoples. The courts must navigate a complicated intersection between the modern regulations and the traditional for environmental justice (Dellapenna & Gupta, 2020). The water resource is a major limitation to growth. It has now been further strained due to climate impact, population growth, and industrial activities. Requesting the legal expert to share their views on intergenerational justice, in light of limited water resources. Using water for AI infrastructure can harm future generations' livelihood. Incorporating sustainability principles in legal frameworks is essential to ensure that future generations will be able to meet their own needs without compromising current choices (Geissdoerfer et al., 2017). Environmental justice is closely related to corporate social responsibility (CSR). The corporate social responsibility of technology companies with AI data centers is to reduce the environmental impact and do no harm to the local community. Corporate social responsibility (CSR) initiatives are becoming almost obligatory standards of conduct, especially in jurisdictions with well-developed environmental governance regimes, due to litigation and regulatory actions.

There are obstacles to achieving environmental justice in AI water use governance. When economic development is prioritized, legal systems can have competing interests that struggle to reconcile. Communities affected by disparate legal capacity, access to justice, and institutional strength cannot challenge negative decisions easily. Anything that takes much time and resources is often hard for marginalized groups to access. There is a need for systemic reforms in order to enhance access to justice, strengthen regulatory oversight, and ensure equitable decision-making processes. As AI infrastructure continues to grow, we will require more integration of environmental justice principles into law and policy. Courts and regulators will set legal precedents for the future in favor of the community and environment.

Corporate Accountability, Transparency, and Regulation

The rapid growth of AI infrastructure is increasing the pressure on water and highlighting the gaps in corporate accountability and regulation. As major consumers of water through data centers, multinational tech firms are facing legal and policy questions about their responsibility, disclosure, and regulation. According to Borgesius et al. (2022), the rising imbalance between business power and regulation capacity lies at the heart of the social media and web-based service space. Big techs have a great deal of money, technical know-how, and operate globally to influence regulation and negotiate on infrastructure. A major issue in corporate water usage is transparency, with many tech firms regarding such data as proprietary information. Nonetheless, there are calls for increased accountability from regulators, civil society, and affected communities. There are lawsuits against corporations for refusing to disclose relevant environmental information that causes significant impacts on nature and people. The gradual shift towards transparency could be regulatory and judicial, with some jurisdictions requiring that companies disclose their use of water metrics (Crawford, 2021). The absence of uniform reporting standards regarding the usage of water across the various jurisdictions poses a challenge around confidentiality and transparency, with implications for comparability, regulatory oversight, and AI infrastructure sustainability assessments. This has increased awareness of the need for international guidelines or harmonized standards for consistent and reliable reporting (IEA, 2022).

Corporate accountability relates to the adequacy of EIAs and compliance regulatory mechanisms. After a detailed evaluation of the literature, researchers conclude that EIA in India may be missing water impacts. Legal issues allege that a lack of proper or no assessment undermines regulatory licenses. Courts consider the cases and enforce the standards of compliance and apply the environmental considerations to the project planning (Geissdoerfer et al, 2017). Corporate environmental responsibility does not just imply merely following the law. IT firms are likely to embrace sustainable practices such as water-efficient cooling technologies, recycling, and other alternative forms of cooling. Although it is often not enforced, lawsuits and government action are usually employed to make companies comply with their own promises (Li et al., 2023). The trends of AI infrastructure development in modern society are increasingly strict regulations. Others have penalty mechanisms, and some jurisdictions have adopted a conditional permitting system, where users must obtain approval for water usage that meets some environmental standards. These innovations will allow adjusting the newly emerged legal tools to digital infrastructure (UNESCO, 2023). Corporate litigation strategy greatly influences the regulatory environment, with tech companies appealing such decisions and negotiating more advantageous terms or postponements of restrictive measures. In countries where the courts view environmental and administrative law, these affect legal precedents. Civil society and other community groups are resorting to litigation to bring corporate players to book. In line with this is the litigation frenzy that results in a process characterized by competing claims and counterclaims.

The integration of environmental, social, and governance principles into the corporate business and regulatory demands is on the increase among investors, regulators, and other stakeholders. This transformation urges organizations to embrace the practice of transparency and sustainability to conform to the laws and regulations, earn a reputation, and secure capital. The effective implementation of the ESG framework can be achieved through high-quality reporting standards and enforcement through legal and regulatory mechanisms (Borgesius et al., 2022). Despite the recent advances, it is still hard to guarantee corporate responsibility with regard to AI water consumption. Standards are complicated by regulatory fragmentation, enforcement capacity disparities, and the computer network nature of AI. In a rapidly changing world, corporate practices can outpace regulations and address them before it is too late. We need stronger legal standards and enforcement mechanisms. Above all, there needs to be a fundamental shift in corporate perception of environmental responsibilities. We will look into future policies and legal pathways.

Future Legal Pathways and Policy Directions

Legal issues around water use in AI infrastructure are fast emerging as a key nexus of global governance. The approaches available currently are reactive and fragmented. Future legal pathways and policy directions may lead to a more coherent and sustainable governance regime for AI-related water usage. In light of rapidly evolving digital technology and its concentration in cities, traditional water law must necessarily evolve from a sectoral regulation approach to an integrated and multi-resource management approach. Policymakers are called upon to re-categorize water use through AI infrastructure and legal classification, as well as to clarify prioritization criteria for water use to eliminate regulatory ambiguity and provide better consistency of decision-making between jurisdictions (Dellapenna & Gupta, 2020). Mandatory transparency and disclosure frameworks are critical for effective governance of water use. New rules are expected to require companies to disclose their water usage, sources of water, and measures taken to improve efficiency. This can form part of more general reporting systems which enhance accountability, allow public participation, and enable comparisons between companies (IEA, 2022). AI infrastructure will likely bring water-specific regulatory instruments, including licensing regimes or sustainability benchmarks, for regulating water extraction. Dynamic allocation mechanisms in the form of seasonal permits could be envisaged for water-distressed regions to allow legal systems to adapt to environmental changes while keeping the technology dynamic. (UNESCO, 2023)

Incorporating precaution and sustainability in legal reasoning is promising due to the uncertainties surrounding the long-term effects of AI infrastructure on the environment. The new legislation reflects the environmental law trend away from traditional pollution attribution. Furthermore, companies internalize their environmental cost and promote development with sustainability principles as a factor (Geissdoerfer et al., 2017). The international and transnational law of water governance will be influenced by the AI infrastructure as it transcends borders. This results in a conflict in the water governance regime and a second challenge of regulatory arbitrage. Establishing baseline standards for environmental performance, transparency, and accountability requires international cooperation; achieving this goal relies on soft-law instruments to harmonize approaches (Borgesius et al., 2022). Future legal frameworks will respect communities' rights and participatory governance mechanisms that create an opportunity for communities to drive decision-making on AI infrastructure. Public participation and legal action are also encouraged, including better access to information, public consultations, and opportunities for those communities affected by water access to sue (Mashaw, 2018). Corporate governance and self-regulation are likely to develop with pressure on IT companies to take conscious steps to sustainability, such as water neutrality, alternate cooling technologies, community partnerships, etc. While voluntary initiatives often do fall short, they may provide a complement to regulation and help create ownership. The sustainability of operations can be ensured through corporate adherence to the law and ethical norms.

While alternative cooling technologies are being pursued, it is equally important that future governance of AI infrastructure adopts a broader sustainability-oriented regulatory framework that brings together legal, technological, environmental, and economic solutions. Governments can establish mandatory water-efficiency requirements, deploy real-time environmental monitoring systems and dynamic permitting, tied to water availability in the region. They can also encourage the adoption of circular water approaches, whereby facilities safely recycle wastewater, harvest rainwater, and use closed-loop cooling systems that dramatically reduce the facility's consumption of fresh water. For those facilities that rely on energy-intensive cooling systems, the incorporation of renewable energy, as well as the deployment of distributed computing across multiple regions, can help to reduce the ecological footprint in water-scarce regions. Future laws and regulations should also require cumulative environmental impact assessments for clusters of hyperscale data centers rather than individual facility-level assessments. Going forward, there will be a strong need for cross-border cooperation around issues of environmental disclosure, sustainability reporting, and corporate accountability. Meanwhile, adequate space for public involvement, community benefit agreements, and environmental justice safeguards and protections are necessary to ensure that facilities do not overconsume resources and are distributed fairly and transparently. Finally, significant investment in green AI research, low-energy computational models, and sustainable digital infrastructure will be needed to reduce water and energy consumption over the long term. Legal adaptation to evolving technologies and environmental conditions holds the key to future AI water governance. Combining new laws, regulations, court interpretations, and major corporations accepting responsibility will be required. Regulatory bodies with specialized knowledge or hybrid governance systems utilizing expertise from environmental science and technology, including law, can help in decision-making. The future of AI water governance will continue to be determined by court judgments in the form of rulings that clarify legal principles, corporate responsibility, and regulatory compliance standards. This approach could entice suitable technological advancement while ensuring sustainability and equity.

Figure 3. Future Sustainable Governance Model for AI Infrastructure

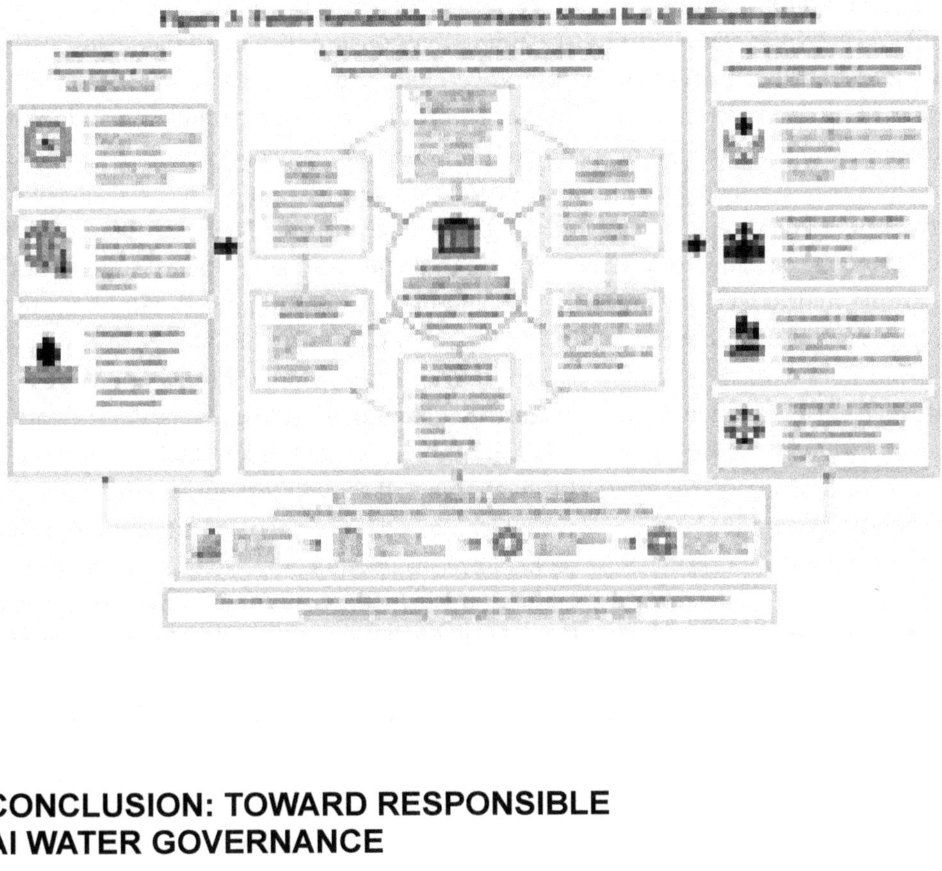

CONCLUSION: TOWARD RESPONSIBLE AI WATER GOVERNANCE

The rapid and smooth momentum of AI infrastructure is focusing attention on the physical aspects of digital transformation, such as water. The problem's statement that started as a technical problem requiring a solution for the optimal use of critical infrastructure has grown and developed over the years into a legal problem or a governance-related problem with serious implications, including harm to environmental sustainability, social equity, and global standards, that is able to affect the world. This Chapter is about a water use lawsuit concerning Artificial Intelligence. The phenomenon that we are observing is not an isolated incident but rather something that reflects some deeper transformations. These are concerned with how technology and its environmental impact are construed and governed. The operations of current laws are already stretched, with landmark instances exemplifying the aforementioned. Laws on environment and water management did not factor in the attributes of AI infrastructure. Judicial and other regulatory frameworks are expected to frequently reinterpret and repurpose pre-existing frameworks towards responding to new issues regarding resources, the firm, and the public interest. New precedents are being set through these litigations, which could guide the future of AI regulation for decades.

As AI gets integrated with physical things that consume finite ecological resources, the study shows the need for technology governance to include sustainability concerns. As an AI program diverts water from a river for a growing farming area, onlookers wonder if water is the sign of justice or power. Such a power is likely to impact an already disadvantaged group, like the rural people and women, more than others. Strict rules should be introduced in the governance structure for it to become more inclusive and participatory. Technology companies are facing increasing scrutiny regarding their environmental behavior and disclosure obligations. However, they have refrained from corporate social responsibility behavior. This would help to align corporate practices with broader interests. The Chapter introduces a variety of pathways to governance of the AI infrastructure, which could enhance classification (in law), require reporting standards, introduce precautionary and sustainability principles, and broaden community engagement. It shows the need for cross-national cooperation and global standards for international oversight of AI infrastructure and its environmental footprint. New Field of Environmental Governance: Regulation of Digital–Ecological Systems. The choices made by the court, regulators, policymakers, and industry actors impact AI development. Those will ensure that there will be the development of AI in a tech-savvy, environment-friendly way in the future. An AI-based water governance framework has been developed to sustain development.

REFERENCES

Arizona v. California, 373 U.S. 546 (1963).

Borgesius, F. Z., Trilling, D., Moeller, J., Bodó, B., de Vreese, C. H., & Helberger, N. (2022). Should we worry about filter bubbles? *Internet Policy Review, 11*(1), 1–16. DOI: 10.14763/2022.1.1561

Bund für Umwelt und Naturschutz Deutschland v. Germany (Weser River Case), Case C-461/13 (2015).

Crawford, K. (2021). *Atlas of AI: Power, politics, and the planetary costs of artificial intelligence*. Yale University Press.

Dellapenna, J. W., & Gupta, J. (2020). *The evolution of the law and politics of water*. Springer.

Food Marketing Institute v. Argus Leader Media, 588 U.S. ___ (2019).

Friends of the Earth, Inc. v. Laidlaw Environmental Services, 528 U.S. 167 (2000).

Geissdoerfer, M., Savaget, P., Bocken, N. M. P., & Hultink, E. J. (2017). The circular economy: A new sustainability paradigm? *Journal of Cleaner Production, 143*, 757–768. DOI: 10.1016/j.jclepro.2016.12.048

Getches, D. H. (2017). *Water law in a nutshell* (5th ed.). West Academic Publishing.

Illinois Central Railroad Co. v. Illinois, 146 U.S. 387 (1892).

International Energy Agency. (2022). *Data centers and data transmission networks*. IEA. https://www.iea.org/

International Energy Agency. (2024). *Electricity 2024: Analysis and forecast to 2026*. IEA.

Li, P., Yang, J., & Mohammad, A. Islam, & Shaolei Ren. (2023). *Making AI less "thirsty": Uncovering and addressing the secret water footprint of AI models*. arXiv. https://doi.org//arXiv.2304.03271DOI: 10.48550

Masanet, E., Shehabi, A., Lei, N., Smith, S., & Koomey, J. (2020). Recalibrating global data center energy-use estimates. *Science, 367*(6481), 984–986. DOI: 10.1126/science.aba3758 PMID: 32108103

Mashaw, J. L. (2018). *Administrative law: The American public law system*. West Academic Publishing.

Mazibuko v. City of Johannesburg, 2009 ZACC 28.

M.C. Mehta v. Kamal Nath, (1997) 1 SCC 388.

Mytton, D. (2021). Data centre water consumption. *npj. NPJ Clean Water*, *4*(11), 1–6. DOI: 10.1038/s41545-021-00101-w

National Audubon Society v. Superior Court, 33 Cal. 3d 419 (1983).

. Rol No. 5888-2019.

Sax, J. L. (1970). The public trust doctrine in natural resource law: Effective judicial intervention. *Michigan Law Review*, *68*(3), 471–566. DOI: 10.2307/1287556

Schlosberg, D. (2013). Theorising environmental justice: The expanding sphere of a discourse. *Environmental Politics*, *22*(1), 37–55. DOI: 10.1080/09644016.2013.755387

Subhash Kumar v. State of Bihar, (1991) 1 SCC 598.

UNESCO. (2023). *United Nations world water development report 2023: Partnerships and cooperation for water*. UNESCO Publishing.

United Nations General Assembly. (2010). The human right to water and sanitation (Vol. A). United Nations., RES/64/292.

Van der Veen, R., & Smith, A. (2022). Cooling the algorithm: The legal and ecological footprint of hyperscale data centers. *Journal of Digital Sustainability*, *5*(2), 88–104.

Vellore Citizens Welfare Forum v. Union of India, (1996) 5 SCC 647.

Waddenvereniging & Vogelbeschermingsvereniging v. Staatssecretaris van Landbouw, Case C-127/02 (2004).

. *Waddenvereniging and Vogelbeschermingsvereniging*, Case C-127/02, ECLI:EU:C:2004:482 (Court of Justice of the European Union 2004).

Wood, M. C. (2014). *Nature's trust: Environmental law for a new ecological age*. Cambridge University Press.

Xiao, T.. (2025). Environmental impact and net-zero pathways for AI infrastructure. *Nature Sustainability*. Advance online publication. DOI: 10.1038/s41893-025-01681-y

KEY TERMS AND DEFINITIONS

AI Infrastructure: The physical and digital systems supporting AI operations, including data centers, cloud computing networks, servers, and cooling technologies.

Artificial Intelligence (AI): Computational systems capable of performing tasks that normally require human intelligence, including machine learning, data processing, prediction, and automated decision-making.

Beneficial Use Doctrine: A principle in water law requiring water resources to be used efficiently, reasonably, and in a socially valuable manner.

Climate Vulnerability: The degree to which communities, ecosystems, or infrastructure are susceptible to adverse impacts of climate change.

Community Rights: Legal and social rights protecting communities' access to natural resources, participation in decision-making, and environmental protection.

Corporate Accountability: The responsibility of corporations to comply with legal, ethical, environmental, and social obligations arising from their operations.

Data Center: A facility that houses computing systems, storage infrastructure, and networking equipment used for processing and managing digital data.

Digital Sustainability: The integration of environmental responsibility and resource efficiency into digital technologies and infrastructure systems.

Environmental Impact Assessment (EIA): A regulatory process used to evaluate the likely environmental effects of a proposed project before approval is granted.

Environmental Justice: The principle that environmental benefits and burdens should be distributed fairly across all communities without discrimination or disproportionate harm.

Environmental Transparency: The disclosure of environmental information, including water consumption and ecological impact, to regulators and the public.

ESG (Environmental, Social, and Governance): A framework used to evaluate corporate performance in environmental responsibility, social impact, and governance practices.

Hyperscale Data Center: A large-scale data center designed to efficiently support massive cloud computing and AI workloads with extensive computing and storage capacity.

Precautionary Principle: A governance principle requiring preventive action where environmental harm is possible, even if complete scientific certainty is unavailable.

Public Trust Doctrine: A legal doctrine stating that certain natural resources, including water, are preserved for public use and must be protected by the state for present and future generations.

Regulatory Arbitrage: The practice of locating operations in jurisdictions with weaker environmental or legal regulations to reduce compliance obligations.

Sustainable Development: Development that satisfies present needs without compromising the ability of future generations to meet their own needs.

Water Governance: The legal, institutional, and policy mechanisms through which water resources are allocated, regulated, managed, and protected.

Water Scarcity: A condition in which available freshwater resources are insufficient to meet environmental, industrial, agricultural, and human demands.

Water-Efficient Cooling Technologies: Cooling systems designed to reduce freshwater consumption in industrial and data center operations.

Chapter 7
Governing AI Data Center Growth in Water-Stressed Regions:
Siting, Transparency, and Water Justice

Mohammad Al Khaldy
http://orcid.org/0009-0009-7502-4668
University of Petra, Amman, Jordan

Youcef Gheraibia
De Montfort University, UK

Ala Hamarsheh
http://orcid.org/0000-0002-4736-7331
Arab American University, Palestine

ABSTRACT

AI data centers are becoming critical infrastructure, but their growth is also a water-governance issue. Cooling systems can raise local freshwater demand directly, while electricity procurement adds indirect water use across power systems. In water-stressed regions, siting decisions affect not only land and grid capacity, but also basin resilience, transparency, and justice. This chapter develops a governance framework for AI data center growth that combines hydrological screening, public disclosure, and water-justice safeguards. It argues that responsible expansion requires place-based regulation that evaluates cumulative withdrawals, drought sensitivity, emergency curtailment, community participation, and enforceable

DOI: 10.4018/979-8-2600-2761-5.ch007

reporting. Drawing on research on data center sustainability, water disclosure, planning, and environmental justice, the chapter proposes policy instruments for governments, utilities, and operators seeking water-resilient digital infrastructure.

1. INTRODUCTION

Artificial intelligence is transforming data centers from background digital utilities into visible public infrastructures. Earlier debates regarding digital infrastructure concentrated primarily on electricity demand, carbon emissions, tax incentives, and regional growth strategies. The current AI expansion adds urgency because frontier-model training, large-scale inference, and high-availability computing clusters are accelerating facility construction, power procurement, and land acquisition. However, the physical expansion of AI infrastructure cannot be reduced to electricity alone. Cooling systems can rely on potable water, reclaimed water, groundwater, or hybrid arrangements, while electricity procurement can embed substantial off-site water consumption across generation portfolios. In water-stressed regions, the result is a governance challenge that is simultaneously hydrological, economic, and political.

The central contribution of this chapter is to reposition AI data center growth as a basin-governance issue rather than merely a private sustainability or engineering issue. Existing research has examined data center efficiency, the electricity-water nexus, corporate water disclosure, planning transparency, and environmental justice. What remains less developed is an integrated decision architecture that connects these literatures within the specific approval logic of AI infrastructure in water-stressed regions. This chapter contributes that missing bridge by showing how hydrological compatibility, public disclosure, and justice safeguards should be linked across screening, permitting, operation, drought response, and periodic review.

The chapter is conceptual rather than empirical, but it is designed to answer concrete policy questions. First, under what hydrological and social conditions is additional AI data center capacity justified in a water-stressed region? Second, what information must be disclosed for regulators and communities to assess direct withdrawals, indirect water dependence, and drought-related consequences in a meaningful way? Third, how should governments respond when a project appears economically attractive yet fails basin-safety or justice criteria? By framing these as guiding questions, the chapter clarifies that the issue is not whether digital infrastructure is valuable in the abstract, but whether specific projects are governable within place-based limits and public legitimacy.

The urgency of these questions is increasing. Recent analysis by the International Energy Agency indicates that data centers accounted for roughly 1.5% of global electricity consumption in 2024, or about 415 TWh, and that demand could rise to around

945 TWh by 2030 in its base case (IEA, 2025). This growth does not automatically translate into uniform water effects, because water outcomes depend on location, cooling strategy, electricity mix, and basin conditions. Yet it does indicate that AI-related infrastructure expansion will increasingly intersect with water allocation, drought planning, and regional resilience. In other words, the planning problem is shifting from whether digital growth exists to how and where it will be governed.

Water stress makes this shift especially consequential. WRI's Aqueduct 4.0 emphasizes that water risk is highly place specific and that screening tools must be augmented by local and regional analysis rather than treated as self-sufficient answers (Kuzma et al., 2023). A facility that appears efficient at the portfolio level may still intensify cumulative basin stress, displace resilience burdens onto households or ecosystems, or receive public subsidies that are opaque to affected communities. For that reason, the relevant governance question is not simply whether an operator can comply with a narrow permit, but whether the project can coexist with drought planning, ecological thresholds, and fair burden sharing.

To address this challenge, the chapter develops an integrated governance framework built on three pillars: basin-aware siting, transparency by design, and water justice. Basin-aware siting requires hydrological screening, cumulative-impact analysis, and adaptive permit conditions rather than parcel-level convenience alone. Transparency by design requires disclosure obligations that are decision relevant, comparable, and enforceable, including clear treatment of direct and indirect water use. Water justice requires procedural, distributive, recognitional, and remedial safeguards so that public participation and burden sharing are not treated as afterthoughts. Together, these pillars provide a more rigorous basis for governing AI data center growth in water-stressed regions.

The remainder of the chapter proceeds in eleven sections. Section 2 clarifies the conceptual scope and explains how the literature was selected and synthesized. Section 3 reviews the relevant literature and identifies the unresolved governance gap. Section 4 presents the integrated governance framework. Sections 5 through 7 translate the framework into siting, disclosure, and justice mechanisms, respectively. Section 8 provides an illustrative scenario showing how the framework operates in practice. Section 9 offers a policy roadmap that differentiates responsibilities among governments, utilities, regulators, and operators. Section 10 identifies future research directions, and Section 11 concludes. A response-to-reviewers appendix is appended to the end of the revised chapter, as requested.

2. CONCEPTUAL SCOPE AND LITERATURE SYNTHESIS APPROACH

This revised chapter follows a structured conceptual-synthesis approach rather than a formal systematic review. The objective is to assemble an interdisciplinary body of work capable of supporting a policy framework for AI data center approvals in water-stressed regions. To do so, the chapter prioritizes literature from six overlapping domains: data center sustainability, AI sustainability, the energy-water nexus, corporate water disclosure, planning transparency, and environmental or water justice. Because AI infrastructure debates have accelerated in recent years, emphasis is placed on peer-reviewed research from roughly 2018 onward, while still retaining foundational studies that define water stress, disclosure, planning legitimacy, and infrastructure justice.

The inclusion logic is thematic and decision oriented. Sources were retained when they directly informed one or more of the following: how data centers consume water on site; how electricity sourcing adds indirect water dependence; how disclosure systems shape accountability; how planning processes structure participation and legitimacy; how water burdens and protections are distributed; and how public institutions translate these concerns into siting, permitting, and operational obligations. High-authority institutional materials, especially from the International Energy Agency and the World Resources Institute, were used selectively where they provided current quantitative baselines or decision-relevant screening tools that peer-reviewed studies alone did not yet synthesize.

The synthesis process involved grouping the literature into recurring themes and then coding those themes into governance functions. For example, work on data center sustainability and the energy-water nexus was mapped primarily to measurement boundaries and siting logic; research on disclosure and transparency was mapped to accountability instruments; and justice scholarship was mapped to procedural, distributive, recognitional, and remedial safeguards. This is not an exhaustive bibliometric review, and it does not claim statistical representativeness. Its contribution is instead analytic integration: it translates fragmented literatures into a coherent framework for public decision making regarding AI data center growth in stressed basins.

Because several reviewer comments emphasized definitional clarity, the chapter also operationalizes key concepts instead of leaving them as abstract labels. Table 1 therefore defines the principal terms used throughout the chapter and identifies how each concept can be recognized in planning or regulatory practice. This operational step matters because many debates over data center sustainability arise from conceptual slippage, particularly when actors switch among the terms water use, withdrawal, consumption, stress, resilience, and justice without specifying their practical meaning. To support operational clarity, Table 1 defines the main concepts employed throughout the chapter and indicates how each concept can be recognized in governance practice.

Table 1. Operational definitions and practical recognition criteria

Concept	Working definition	How it is recognized in practice	Governance relevance
AI data center	A facility or campus providing large-scale compute, storage, and networking capacity for AI training, inference, and related digital services.	High-density server loads, large power interconnections, dedicated cooling infrastructure, and high-availability operation.	Frames the object of governance as physical infrastructure rather than an immaterial cloud service.
Water-stressed region	A basin or service area where scarcity, variability, ecological limits, or allocation pressure make additional withdrawals socially or environmentally sensitive.	Recurring drought rules, high baseline stress, aquifer decline, ecological flow constraints, or chronic supply-demand imbalance.	Determines whether ordinary permitting is sufficient or enhanced review is required.
Basin-aware siting	A site-selection approach that evaluates the hydrological, social, and infrastructural context of the basin before approval is granted.	Screening based on basin stress, source suitability, cumulative demand, drought triggers, and community vulnerability.	Prevents parcel-level approval logic from overlooking basin-wide risk.
Direct water use	Water withdrawn, recycled, discharged, or consumed on site for cooling, humidification, and related facility operations.	Metered facility withdrawals, seasonal consumption profiles, source-specific reporting, and operating permits.	Supports caps, monitoring, and drought-response conditions.
Indirect water use	Water embodied in electricity generation or other upstream services required to operate the facility.	Electricity-sourcing disclosures, regional generation portfolios, and scenario-based contextual reporting.	Broadens accountability beyond on-site metrics, but should not automatically replace basin-level direct-use controls.

continued on following page

Table 1. Continued

Concept	Working definition	How it is recognized in practice	Governance relevance
Transparency by design	Disclosure rules built into approvals and operations from the start rather than added after controversy emerges.	Application-stage disclosure, periodic public reporting, prompt incident notices, and public dashboards.	Turns sustainability claims into comparable and reviewable information.
Water justice	A governance approach that evaluates participation, burden sharing, recognition, and remedy in water-related decisions.	Early engagement, plain-language disclosure, benefit-burden analysis, curtailment rules, and remedies for non-compliance.	Ensures legitimacy is judged by fairness as well as efficiency or growth.

3. LITERATURE SYNTHESIS AND THE UNRESOLVED GOVERNANCE GAP

Research on digital infrastructure sustainability increasingly demonstrates that data centers cannot be understood through electricity metrics alone. Masanet et al. (2020) showed that data center energy demand does not rise linearly with digital-service growth because efficiency improvements matter, but that result also revealed the importance of system boundaries in sustainability assessment. Mytton (2021) extended the analysis by focusing specifically on water consumption in data centers, thereby highlighting how water accounting remains less mature than energy accounting. More recent studies on AI sustainability further demonstrate that model size, retraining frequency, hardware choice, utilization, and geographic location all affect the environmental profile of computation (Strubell et al., 2019; Lannelongue et al., 2021; Verdecchia, Sallou, & Cruz, 2023; Bolón-Canedo et al., 2024; Li, Yang, Islam, & Ren, 2025).

This literature is valuable because it dislodges the assumption that digital infrastructure is environmentally negligible. Yet it remains only partially adequate for public governance. Much of the sustainability discussion still centers on operator performance, engineering optimization, or portfolio reporting. These perspectives are useful for internal management, but they do not fully answer public questions regarding basin compatibility, cumulative withdrawals, curtailment priorities during drought, or the legitimacy of locating large water-dependent facilities in already stressed regions. A city or basin authority requires more than firm-level efficiency claims; it requires a decision architecture that connects those claims to local hydrology, public oversight, and distributive consequences.

A second body of literature examines the electricity-water nexus. Blackhurst, Hendrickson, and Vidal (2010), Macknick et al. (2012), Stillwell and Webber (2014), Lee et al. (2018), Jin et al. (2019), and Talati et al. (2016) show that electricity generation varies sharply in its withdrawal and consumption profiles depending on generation technology, cooling method, location, and climate. For AI infrastructure, this means that a facility with apparently low on-site water consumption may still depend on a water-intensive electricity portfolio. Farfan and Lohrmann (2023) similarly estimate that digital data services carry both electricity and water footprints that can be understated when analysis is limited to the building envelope. These studies make a decisive point: direct and indirect water accounting serve different governance purposes and should not be conflated.

The key governance challenge is not whether indirect water matters; it clearly does. The challenge is how it should matter in approval and oversight. Some policy instruments should treat direct water withdrawals as mandatory, permit-level control variables because they affect local basins immediately and measurably. Indirect water dependence, by contrast, is usually better treated as a contextual indicator and planning variable unless a jurisdiction has authority to condition electricity sourcing directly. Failure to distinguish these roles leads to confusion. It can also permit rhetorical substitution, where low on-site withdrawals are used to obscure high off-site water dependence, or vice versa.

Corporate water disclosure research provides another critical lens. Ben-Amar and Chelli (2018), Burritt, Christ, and Omori (2016), Yu, Kuo, and Ma (2020), Zhang, Tang, and Huang (2021), Wahyuningrum et al. (2023), Pinheiro, do Prado, and de Moraes (2024), and Farooq et al. (2025) collectively show that water disclosure is shaped by institutional pressure, governance quality, and reporting norms, but that the depth and comparability of disclosure remain inconsistent. Voluntary disclosures may emphasize favorable narratives, aggregate metrics, or enterprise-level targets while underreporting facility-level variation, local stress conditions, or enforcement consequences. For AI data centers, this limitation is severe because local communities and regulators need site-level information rather than high-level sustainability branding.

Planning and transparency scholarship reaches a similar conclusion from a different angle. Marantz and Ulibarri (2022) argue that transparency is not a simple virtue but a tension-filled value: information can be technically disclosed while remaining too aggregated, too selective, or too late to alter outcomes. Wang, Ulibarri, Scott, and Davis (2023) further show that infrastructure and environmental review processes can reproduce inequality when formal participation is not matched by intelligible information and meaningful decision authority. Applied to AI data center siting, this means that disclosure must be judged not by whether a document exists, but by whether it is comparable, understandable, and linked to enforceable decisions.

Environmental and water justice scholarship adds the normative dimension that operator-centered sustainability often lacks. Shrimpton, Hunt, and Rogers (2021), Hoover et al. (2021), Cannon et al. (2024), McKee (2025), and Shahrad, Stålhammar, and Gulsrud (2025) emphasize that just governance requires attention to procedure, distribution, recognition, and remedy. In water-stressed regions, these dimensions become especially salient because scarcity decisions affect households, ecosystems, agriculture, small businesses, and industrial users unevenly. A project may be efficient in engineering terms yet unjust in planning terms if communities are engaged too late, if drought burdens are shifted onto less powerful users, or if benefit-sharing is symbolic rather than substantive.

The AI ethics literature reinforces a related point about governance design. Floridi et al. (2018), Jobin, Ienca, and Vayena (2019), and Mittelstadt (2019) caution that high-level principles are useful but insufficient unless they are translated into measurable obligations, clear authority, and revisable accountability mechanisms. This observation applies directly to AI data center water governance. Principles such as sustainability, stewardship, or justice are politically attractive, but they do not by themselves determine which projects should be denied, resized, phased, or conditioned under drought-sensitive rules. Public institutions require a stepwise pathway that moves from normative commitments to implementation mechanisms.

The literature therefore converges on a governance gap. Data center sustainability research highlights material resource demands but often remains firm centered. Energy-water nexus research clarifies indirect dependencies but rarely translates them into digital-infrastructure approval rules. Corporate disclosure research shows the limitations of voluntary reporting but does not specify basin-sensitive disclosure architectures. Planning and justice research explains legitimacy and burden sharing but is less often connected to the technical specifics of cooling, electricity sourcing, or drought contingencies. The unresolved task is to assemble these insights into a single decision chain for AI infrastructure in water-stressed regions. That is the specific gap this chapter addresses. Because several review comments requested stronger treatment of measurement boundaries, Table 2 clarifies how different water-accounting elements should function in governance rather than leaving them conceptually interchangeable.

Table 2. Measurement boundaries and governance roles for water accounting

Accounting element	What is measured	Recommended governance role	Why it matters	What it should not be used for
Direct water withdrawals/ consumption	On-site metered withdrawals, recycling, discharge, and evaporative losses.	Mandatory disclosure and permit control variable.	Represents immediate basin-level impact and drought sensitivity.	It should not be replaced by portfolio averages or enterprise-level narratives.
Indirect electricity-related water dependence	Water embodied in the electricity portfolio serving the facility.	Contextual indicator and planning variable; mandatory disclosure where feasible.	Prevents narrow reporting that hides upstream water dependence.	It should not automatically function as a permit threshold if regulators lack authority over generation portfolios.
Cumulative basin demand	Interaction of current and projected users within the same basin or service area.	Strategic-screening and permit-review variable.	Identifies cluster effects and long-run allocation pressure.	It should not be approximated only from parcel-level demand.
Drought-stage performance	Facility behavior under heat extremes, shortage declarations, or emergency curtailment.	Adaptive permit trigger and public accountability metric.	Determines whether the facility can coexist with scarcity management.	It should not be left undefined until crisis conditions emerge.
Community burden and benefit profile	Who receives jobs, tax benefits, reliability protections, and who absorbs water-system risk.	Justice-screening and mitigation variable.	Connects technical approval to legitimacy and fairness.	It should not be substituted with generic philanthropic commitments.

4. AN INTEGRATED GOVERNANCE FRAMEWORK

The integrated framework proposed in this chapter treats AI data center governance as a chain of linked public decisions rather than a one-time siting exercise. The chain begins before a parcel is marketed for development, continues through utility negotiation and permitting, extends into routine operation and drought response, and culminates in periodic review. At each stage, three questions must be addressed. Is the proposed growth hydrologically compatible with the basin? Is the relevant information visible, intelligible, and enforceable? Are the burdens and protections distributed justly? If any of these questions is ignored, an apparently efficient project can still generate weak legitimacy and fragile long-term outcomes.

This framework differs from operator-centered sustainability models in two ways. First, it is explicitly place based. Portfolio-level targets are insufficient when water impacts are experienced through local basins, utility systems, ecological thresholds, and community histories. Second, it is authority sensitive. Voluntary initiatives can encourage improvement, but they cannot substitute for enforceable rules on withdrawals, source hierarchies, drought participation, incident disclosure, or adaptive permit review. In water-stressed regions, the governance challenge is not only to encourage better corporate behavior, but to define the public conditions under which infrastructure growth is acceptable at all.

Figure 1 summarizes the chapter's core argument. The framework is organized around three pillars—basin-aware siting, transparency by design, and water justice—that converge in adaptive oversight. Adaptive oversight is necessary because basin conditions, AI workloads, cooling practices, electricity sourcing, and public expectations all evolve over time. A static permit issued under one set of assumptions can become misaligned with later drought cycles, clustered development, or operational changes. By embedding review triggers and public accountability mechanisms, the framework keeps AI infrastructure governance responsive rather than merely reactive.

Figure 1. Integrated governance architecture for AI data center growth in water-stressed regions.

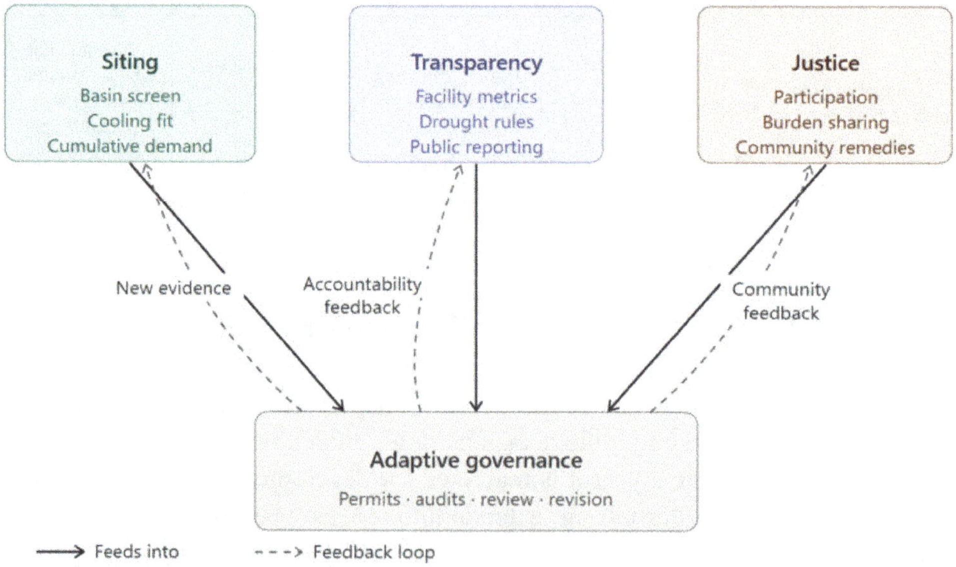

5. BASIN-AWARE SITING AND ADAPTIVE PERMITTING

Siting is the front line of water governance because it determines which hydrological system, community, utility regime, and climate risks a facility will enter. Conventional economic-development logic prioritizes land availability, transmission access, fiber connectivity, construction speed, and tax incentives. Those variables matter, but in water-stressed regions they are insufficient. A basin-aware siting process begins with the question of compatibility rather than mere feasibility. Some basins should be presumptively unsuitable for water-intensive expansion because of chronic scarcity, heavy groundwater overdraft, ecological fragility, or acute inequality in water service. Other basins may be treatable as caution zones in which approval is possible only under stronger proof, tighter operating constraints, and more demanding public oversight.

A robust siting screen must evaluate direct, indirect, and cumulative water effects together. Direct effects include source choice, withdrawals, consumption, discharge, and the reliability of recycled-water infrastructure. Indirect effects include electricity-related water dependence across the supplying generation mix. Cumulative effects include the interaction of multiple industrial users, urban growth, climate volatility, and existing drought rules. This cumulative perspective is essential because one facility may appear manageable in isolation while a cluster of projects becomes politically and hydrologically destabilizing. The relevant unit of analysis is therefore not just the parcel or plant, but the basin and its surrounding infrastructure system.

Cooling strategy should be assessed through scenario analysis rather than annual averages alone. Operators often emphasize average water-use effectiveness or design efficiency. Yet what matters in stressed regions is how the configuration behaves under extreme heat, low reservoir levels, reclaimed-water interruptions, grid constraints, and emergency shortage declarations. A design that appears efficient in average conditions may shift abruptly toward higher freshwater demand when temperatures spike or when a recycled-water feed becomes unavailable. Regulators should therefore require seasonal operating profiles, heat-stress testing, and drought-contingency logic as part of the siting review.

Adaptive permitting follows directly from this logic. A data center approved under one set of basin conditions may become inappropriate as drought deepens, groundwater declines, or cumulative industrial demand rises. Permits should therefore include review triggers linked to material changes in water stress, electricity sourcing, operating assumptions, and regional demand. Adaptive review does not create uncertainty for its own sake. It recognizes that hydrological compatibility is dynamic and that public authorities need lawful mechanisms to revisit conditions when the real world changes. This is especially important for AI infrastructure because demand trajectories can shift faster than many conventional infrastructure forecasts anticipate.

Public participation is particularly consequential during siting because the most path-dependent decisions are made early. If land options are secured, utility extensions negotiated, and incentive packages agreed before communities are informed, later hearings become largely symbolic. A justice-oriented siting process reverses that sequence. Communities should be notified before key commitments are locked in, and engagement should focus on the decisions that materially shape water outcomes: source hierarchy, cooling strategy, curtailment obligations, infrastructure financing, and monitoring arrangements. Consultation that occurs only after these questions have been effectively settled does not satisfy procedural legitimacy.

Economic-development claims should also be tested against hydrological realities. Data center proponents often promise construction activity, tax revenue, strategic prestige, and spillover innovation benefits. Some of these gains may be real, but they are rarely distributed in the same way that water burdens are distributed. A strong siting process should therefore compare projected public benefits with explicit water-system costs, reliability risks, basin constraints, and drought governance implications. This is not anti-growth. It is a way of asking whether digital growth is socially warranted rather than simply politically attractive. Table 3 translates the siting discussion into review criteria that public authorities can apply before a project is approved. Figure 2 presents the adaptive siting workflow discussed above, showing how regional screening, enhanced review, permitting, and reopener mechanisms should connect over time.

Table 3. Basin-aware siting criteria for AI data centers

Criterion	Illustrative questions	Evidence required	Possible decision consequence
Basin stress	Is the watershed already highly stressed, drought prone, or hydrologically volatile?	Hydrological status, stress indicators, drought history, ecological thresholds.	Can trigger exclusion, caution-zone status, or enhanced review.
Water-source fit	Will the facility rely on potable water, groundwater, reclaimed water, or a mixed hierarchy?	Source plan, infrastructure maps, reliability assessment, competing uses.	Shapes source restrictions and permit conditions.
Cooling suitability	How does the chosen cooling configuration perform during heat extremes and shortage periods?	Seasonal profiles, scenario tests, emergency operating logic.	Determines technology approval and contingency obligations.

continued on following page

Table 3. Continued

Criterion	Illustrative questions	Evidence required	Possible decision consequence
Indirect water dependence	How water intensive is the electricity portfolio that will serve the facility?	Electricity-sourcing disclosure, regional generation profile, scenario assumptions.	Adds contextual scrutiny and can support location or procurement conditions.
Cumulative demand	What other large users or likely future projects affect the same basin?	Regional demand forecasts, cluster analysis, utility planning documents.	Can justify phasing, caps, or denial.
Social vulnerability	Which communities or ecosystems would face increased strain, cost pressure, or reduced resilience?	Equity screening, service-quality data, affordability indicators, local testimony.	Requires mitigation, redesign, or stronger review.
Adaptive compatibility	How will the project be revisited if basin conditions materially worsen?	Permit review triggers, monitoring plans, reopener clauses.	Converts approval from static authorization to living governance.

Figure 2. Adaptive siting workflow from regional screening to periodic review

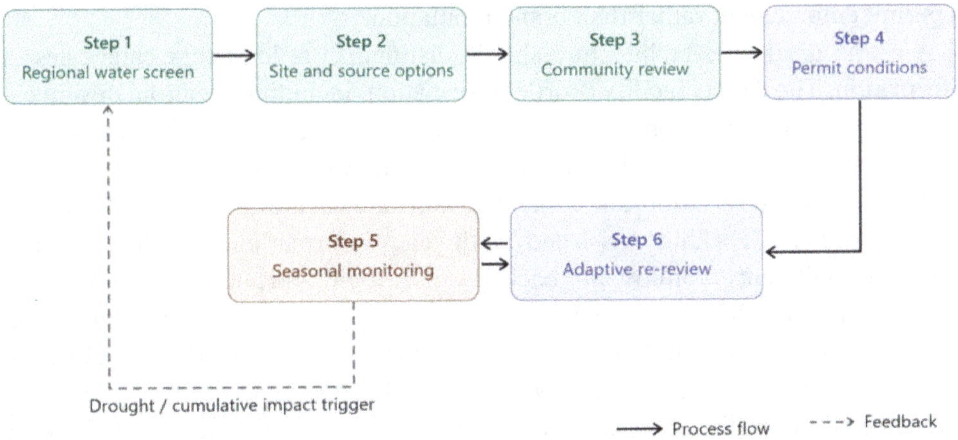

6. TRANSPARENCY BY DESIGN AND MEASUREMENT BOUNDARIES

Transparency is often invoked as a cure for controversy, but meaningful water governance requires a more demanding standard. The relevant question is not whether operators publish sustainability statements. It is whether regulators, utilities, communities, and researchers receive the information necessary to evaluate direct withdrawals, indirect water dependence, cumulative basin effects, and drought behavior in a form that supports oversight. For AI data centers, transparency must therefore be designed around decisions rather than public relations. It should illuminate how much water a facility uses, where that water comes from, what is recycled or discharged, how performance varies seasonally, how electricity sourcing shapes indirect dependence, and what happens under shortage conditions.

The disclosure literature suggests that voluntary reporting rarely satisfies these requirements. Firms disclose more when institutional pressure is stronger, but disclosure quality and comparability vary widely. Enterprise-level sustainability narratives can be polished yet still omit facility-level intensity, seasonal peaks, basin conditions, public subsidies, drought exemptions, or compliance failures. Water-stressed regions cannot rely on selective corporate reporting as the basis for public approval. They require a disclosure architecture that is legally specified, time sensitive, and linked to permit enforcement rather than brand reputation.

A useful disclosure architecture should distinguish at least four categories of information. The first is facility profile information, including location, design capacity, cooling configuration, source hierarchy, and expected load profile. The second is water accounting information, including withdrawals, consumption, recycling, discharge, seasonal variation, and contextual indicators of indirect electricity-related water dependence. The third is risk and contingency information, including shortage triggers, curtailment commitments, backup supply logic, and public communication protocols. The fourth is governance information, including permit conditions, audit results, community commitments, grievance channels, and incident disclosures. These categories matter because they connect technical metrics to accountability.

Scale and timing are central. Annual enterprise-wide disclosures are too coarse for stressed basins where seasonality, drought declarations, and local peaks determine legitimacy. Facility-level reporting should be the default, with at least seasonal reporting for direct water metrics and prompt public notice when material deviations, outages, exceedances, or emergency exemptions occur. Indirect water dependence should also be disclosed, but with clear explanation of its role. As argued earlier, indirect water metrics are most useful as contextual and comparative indicators unless jurisdictions have the legal capacity to condition electricity portfolios directly.

The chapter therefore proposes mandatory disclosure of indirect dependence, but not automatic use of that metric as a standalone denial threshold.

Transparency should also extend to the public side of the bargain. Local governments and utilities often provide infrastructure upgrades, tariff arrangements, land-use accommodations, or expedited approvals to attract large projects. If these arrangements remain opaque, the public cannot determine whether scarce water-system capacity is being subsidized, reprioritized, or protected on behalf of a single operator. Water-transparent governance therefore requires disclosure of material utility commitments, infrastructure financing arrangements, drought exceptions, and benefit agreements. Without that visibility, the formal availability of sustainability data may coexist with real opacity about how public resources are being allocated.

Importantly, transparency is not an end in itself. Its function is to support comparison, contestation, enforcement, and learning. Data should therefore be disclosed in formats that can be benchmarked across facilities and over time, not buried in promotional summaries or highly technical appendices. Public dashboards, independent audits, plain-language summaries, and explanation requirements can help translate raw numbers into accountability. Done well, transparency changes incentives by making overstatement, selective disclosure, and vague commitments harder to sustain. Done poorly, it can create only the appearance of openness. The disclosure architecture summarized in Table 4 specifies the minimum information that should be publicly available, and the purpose served by each disclosure category. Figure 3 visualizes the transparency stack proposed in this section, linking facility metrics to contextual interpretation, public visibility, and corrective action.

Table 4. Disclosure architecture for water-transparent AI data centers

Disclosure area	Minimum content	Recommended cadence	Primary accountability value
Facility profile	Location, design capacity, cooling system, source hierarchy, expected load trajectory.	At application and whenever materially changed.	Allows the public to understand what is being approved and operated.
Direct water metrics	Withdrawals, consumption, recycling, discharge, seasonal variation, source-specific breakdown.	Monthly or seasonal, with annual summary.	Supports monitoring against permit assumptions and drought obligations.
Indirect water metrics	Electricity-related water dependence, sourcing logic, material changes in procurement assumptions.	Annual with material updates.	Broadens accountability beyond on-site use and prevents narrow framing.

continued on following page

Table 4. Continued

Disclosure area	Minimum content	Recommended cadence	Primary accountability value
Drought and contingency rules	Curtailment triggers, emergency exemptions, backup logic, communication protocol.	Before operation and promptly when changed.	Clarifies scarcity governance before crisis conditions arise.
Compliance and incidents	Permit exceedances, reporting failures, outages, audit results, emergency departures from plan.	Prompt public notice and periodic reporting.	Transforms disclosure into enforceable oversight.
Public commitments	Community benefits, infrastructure financing, monitoring obligations, grievance channels.	At agreement signing and during periodic review.	Shows whether the public bargain is being honored.

Figure 3. Transparency and accountability stack for AI data center water governance.

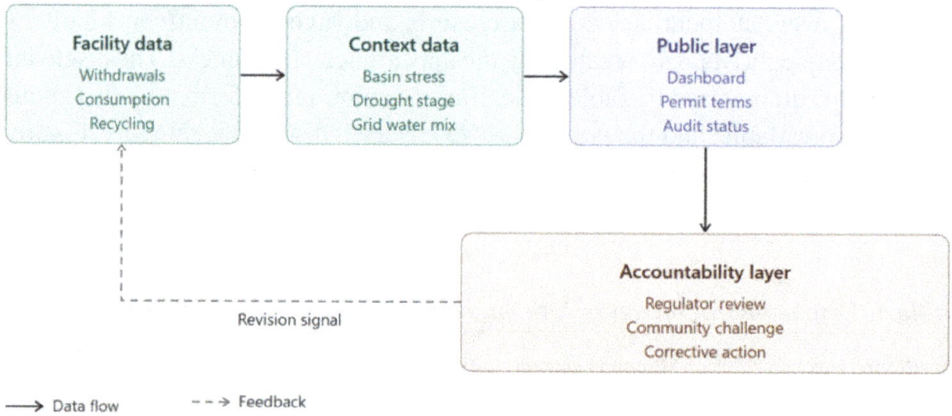

7. WATER JUSTICE, TRADE-OFFS, AND DECISION RULES

Water justice broadens the governance lens beyond efficiency. A facility may display respectable water-use indicators and still be unjust if communities are engaged too late, if drought burdens fall disproportionately on low-income households, or if public resources are redirected toward industrial reliability while ordinary service quality deteriorates. Justice scholarship is therefore indispensable for AI data center governance because it clarifies that legitimacy depends on how decisions are made, how burdens are shared, whose knowledge counts, and what remedies exist when commitments fail.

Procedural justice requires more than statutory notice. Communities need early engagement, intelligible information, and meaningful opportunities to influence the terms of approval before those terms become path dependent. A hearing held after utility commitments and incentive packages are effectively fixed does not create real participation, even if the formal process appears compliant. Procedural justice also requires support for participation. Large infrastructure applicants can afford consultants, legal teams, and complex scenario modeling; community groups often cannot. Governments should therefore consider funding independent technical review or community expert support when major water-dependent AI projects are proposed.

Distributive justice asks how benefits and burdens are allocated. AI data centers are frequently justified in terms of construction jobs, tax receipts, and innovation branding. Some of these benefits are real, yet they may be temporary, geographically diffuse, or inaccessible to the households that bear most of the water-related risk. Water burdens, by contrast, are often immediate and local: supply competition, affordability concerns, infrastructure strain, and uncertainty during drought. A just review should therefore compare who receives durable benefits with who absorbs durable risks. Where these distributions diverge, mitigation, redesign, phasing, or denial should remain live options rather than symbolic add-ons.

Recognitional justice highlights histories that conventional cost-benefit analysis often ignores. Communities may have prior experience with industrial siting, service inequities, or environmental sacrifice that shapes how they interpret new water-intensive projects. Ignoring those histories can make even technically polished reviews substantively unjust. Recognition means taking local experience seriously, providing language access and accessible meeting formats, and acknowledging that formally equal participation may still reproduce inequality when actors differ sharply in resources and institutional access.

Remedial justice is equally important. Many project controversies persist because commitments are framed as aspirations rather than obligations. A community benefit fund or sustainability pledge may sound constructive but remain trivial relative to basin risk if no measurable standard, audit path, or revision mechanism is attached. The chapter therefore treats remedy as a core justice dimension: permits and agreements should specify what happens if promised performance, disclosure, curtailment, or community commitments are not met. Without credible remedies, justice language can become a form of legitimacy management rather than a substantive governance principle.

Trade-offs among equity, efficiency, and development goals must be confronted explicitly. A project may promise economic benefits yet still fail basin-safety or justice criteria. In such cases, the appropriate response should not be rhetorical balancing that treats all objectives as commensurable by default. Instead, the governance system should establish priority rules. Projects that cannot operate within hydrological limits or that depend on opaque or inequitable drought privileges should be denied, resized, phased, or redesigned before economic benefits are weighed as decisive. Efficiency gains cannot legitimize projects that externalize water insecurity onto less powerful users. Conversely, where basin compatibility can be demonstrated and justice concerns can be mitigated credibly, approval may be justified under strong transparency and adaptive-review conditions. Table 5 converts the justice principles in this section into screening questions that can be used during approvals, renewals, and drought-related review. Figure 4 is placed here because it shows how procedural, distributive, recognitional, and remedial justice interact with adaptive permitting and social license.

Table 5. Water-justice screening questions for approvals and renewals

Justice dimension	Screening question	Risk signal if unanswered	Typical governance response
Procedural	Were affected communities informed before key commitments became locked in?	Late notice and symbolic consultation.	Reopen review, extend engagement, and fund independent support.
Procedural	Can non-experts understand the project's water assumptions and drought rules?	Technical opacity and unequal participation.	Require plain-language summaries and public explanation.
Distributive	Who receives durable benefits and who bears water-system risks?	Benefit-burden mismatch.	Condition approval on mitigation, redesign, or revised terms.
Distributive	Who curtails first during shortage or emergency?	Ad hoc prioritization favoring powerful users.	Define ex ante curtailment hierarchy in permits and utility agreements.
Recognitional	Were local histories of exclusion, affordability stress, or service inequality acknowledged?	Formal equality masking substantive inequality.	Incorporate history into impact review and benefit design.
Remedial	What remedies exist if promised performance or community commitments fail?	No credible accountability.	Require audits, penalties, permit reopeners, and revisable approvals.

Figure 4. Water-justice governance loop for AI data center approvals.

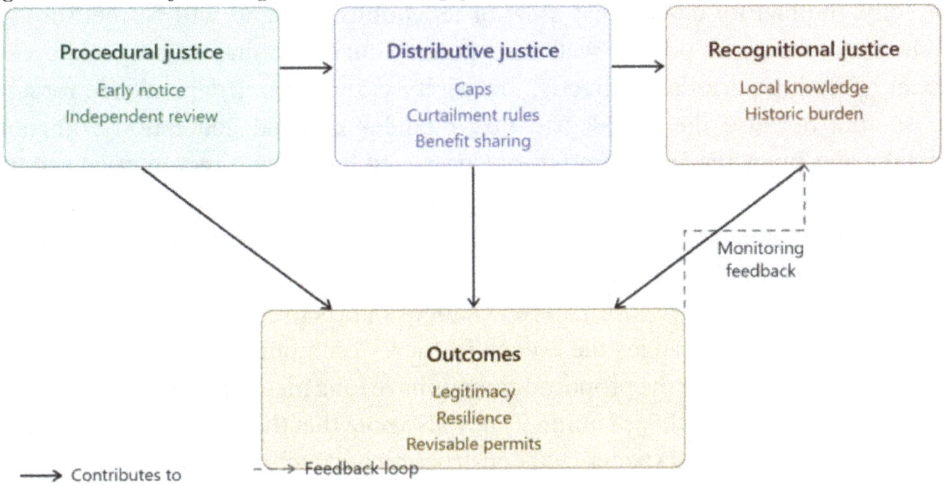

8. ILLUSTRATIVE SCENARIO: REVIEWING A PROPOSED AI CAMPUS IN A SEMI-ARID BASIN

To make the framework more concrete, this section presents an illustrative scenario rather than a claimed empirical case study. Imagine a fast-growing metropolitan region in a semi-arid basin where municipal leaders are actively recruiting digital infrastructure. The basin already experiences seasonal restrictions, rising temperatures, groundwater pressure in surrounding agricultural districts, and uneven service reliability in several lower-income neighborhoods. A major operator proposes a 600-MW AI data center campus that would be developed in phases, using a hybrid cooling scheme, a reclaimed-water agreement with the city utility, and a portfolio of grid and power-purchase electricity arrangements.

Under a conventional development model, the project might advance quickly because the site has land, transmission potential, and strong political support. Under the framework proposed in this chapter, however, the first question is compatibility. Basin screening reveals that the region qualifies as a caution zone rather than a preferred zone because seasonal stress is high, reclaimed-water infrastructure is not yet fully reliable in summer peaks, and two other industrial projects are already in late-stage negotiation. The project is therefore not denied outright, but it is shifted into enhanced review. That enhanced review requires a basin-sensitive water statement, scenario-based testing of the cooling design, disclosure of the source hierarchy, and explicit drought-curtailment commitments before utility contracts are finalized.

The siting review then distinguishes direct from indirect water dependence. Directly, the facility claims that most of its cooling demand will be met through reclaimed water, with potable water as a backup during commissioning and exceptional operating periods. Indirectly, electricity-related water dependence remains substantial because the initial grid mix includes thermal generation with non-trivial water intensity. In the framework proposed here, the direct metrics become permit-control variables, while the indirect metrics become mandatory contextual disclosures. Regulators do not deny the project solely because its off-site electricity carries water dependence, but they require the operator to disclose the magnitude of that dependence and to report material changes in procurement strategy over time.

Justice screening changes the review further. Community groups point out that two neighborhoods near the proposed campus have long histories of service interruptions and low trust in utility planning. They also note that the promised employment profile is skewed toward specialized construction and technical roles unlikely to be broadly accessible without prior workforce investment. Under a justice-by-design approach, these concerns are not relegated to a generic community-benefit discussion. Instead, they become part of the approval test. The operator is required to fund independent technical support for community review, participate in a public explanation process regarding water assumptions, and negotiate measurable commitments that directly strengthen local water resilience rather than relying solely on broad philanthropic pledges.

The drought-curtailment question is decisive. The operator initially seeks broad flexibility, arguing that the campus is strategically important and should retain high availability during heat waves. The framework proposed in this chapter rejects undefined scarcity privileges. Utility and regulatory approvals therefore require an ex ante curtailment hierarchy. During specified drought stages, the operator must reduce direct freshwater fallback use, suspend non-essential expansion activities, and accept phased curtailment conditions unless public health or grid-reliability exceptions are invoked through a transparent process. The operator may continue to rely on reclaimed water where safe and contractually secure, but the permit makes clear that drought governance is a public rule rather than a private negotiation in crisis.

Periodic review then closes the loop. If the campus expands, if reclaimed-water reliability changes, if clustered development intensifies regional demand, or if reported performance diverges materially from the approval assumptions, the permit is reopened. Community oversight meetings and public dashboards allow stakeholders to track whether the project's water bargain is being honored. In this scenario, the framework does not prohibit digital development categorically. Rather, it makes social license contingent on hydrological compatibility, measurable transparency, and justice safeguards. The result is a more demanding but more durable model of approval.

9. POLICY ROADMAP AND ACTOR RESPONSIBILITIES

The revised framework becomes most actionable when responsibilities are differentiated clearly across institutions. Governments, utilities, regulators, and operators all influence project outcomes, but they do not control the same instruments or face the same accountability tests. One recurring weakness in current practice is that duties are blurred. Water sourcing may be treated as a utility matter, disclosure as a sustainability matter, justice as a planning matter, and drought behavior as an emergency matter, leaving no actor clearly responsible for the whole governance picture. The policy roadmap below addresses this fragmentation by assigning duties across the project life cycle.

Regional and local governments should lead strategic planning. Their role is to establish exclusion zones, caution zones, basin-sensitive screening standards, and the public conditions under which digital infrastructure recruitment is appropriate. They also control many of the incentives that shape project pipelines. If economic-development subsidies are offered without water-sensitive conditions, public finance can unintentionally reward hydrologically fragile growth. Governments should therefore condition major incentives, land-use facilitation, or accelerated review on basin compatibility, facility-level disclosure, drought participation, and community-accountability commitments.

Utilities are not passive service providers in this framework. Water and electricity utilities can require large digital customers to disclose operational assumptions, participate in contingency planning, finance connection-specific infrastructure, and accept curtailment obligations under defined conditions. Where reclaimed water is proposed, utilities should evaluate reliability and competing uses honestly rather than presenting recycled supply as a universal solution. Where electricity-water dependence is material, utilities and energy regulators should coordinate so that local siting decisions do not rest on narrow assumptions about upstream resource availability.

Environmental and planning regulators are responsible for converting approval logic into enforceable obligations. They should require basin-sensitive water statements at the application stage, specify source hierarchies and reporting cadence in permits, establish incident-notice requirements, and define adaptive review triggers. Regulators also have a particular responsibility to ensure that disclosure remains intelligible rather than merely voluminous. Plain-language public summaries, consistent reporting templates, and publicly accessible dashboards help make regulatory information usable outside specialist circles.

Operators, finally, must shift from reputational sustainability to auditable stewardship. This means embedding facility-level metering, seasonal reporting, independent verification, internal governance records, and public grievance mechanisms into routine management rather than activating them only when controversy arises. It also means recognizing that social license in stressed basins is not secured by branding alone. An operator that treats transparency, drought participation, and community commitments as core operating disciplines will be more resilient over time than one that seeks flexibility through opacity. Table 6 differentiates primary responsibilities across the main actors involved in project review and operation, while Table 7 complements the actor matrix by locating the main policy instruments at each stage of the project life cycle. Figure 5 summarizes how actor responsibilities shift over the project life cycle and why coordination failures can undermine otherwise strong individual instruments.

Table 6. Actor-responsibility matrix for AI data center water governance

Actor	Primary responsibilities	Typical timeline	Key accountability instruments
Regional/local governments	Map exclusion and caution zones; align economic-development incentives with water goals; ensure early public engagement.	Strategic planning and pre-application stages.	Zoning tools, incentive conditions, public-hearing requirements, intergovernmental agreements.
Water utilities	Evaluate source hierarchy; assess reclaimed-water reliability; negotiate contingency and curtailment terms; publish material service commitments.	Pre-application, permitting, and drought periods.	Service agreements, drought-stage rules, infrastructure financing conditions, public dashboards.
Electricity utilities / energy regulators	Disclose relevant electricity-sourcing assumptions and local grid implications; coordinate where upstream water dependence is material.	Pre-application, procurement changes, periodic review.	Grid-impact assessments, procurement disclosure, coordination protocols.
Environmental/planning regulators	Require basin-sensitive water statements; set permit conditions; enforce monitoring, incident notice, and adaptive review.	Application, permitting, operation, and renewal.	Permit clauses, audits, penalties, permit reopeners, disclosure templates.

continued on following page

Table 6. Continued

Actor	Primary responsibilities	Typical timeline	Key accountability instruments
Operators	Provide accurate facility-level data; implement approved source hierarchy and drought behavior; fund monitoring and community commitments.	Application through operation.	Metering, audit trails, reporting systems, community-benefit agreements, grievance channels.
Community oversight bodies / civil society	Review disclosures; raise local knowledge and burden-sharing concerns; monitor whether commitments are honored.	Throughout project life cycle.	Oversight committees, public comment, citizen monitoring, independent technical review.

Table 7. Policy instruments across the AI data center project life cycle

Project stage	Lead actors	Illustrative instruments	Primary purpose
Strategic planning	Regional and local governments	Exclusion zones, caution zones, basin maps, cumulative-demand thresholds.	Prevent unsuitable projects from advancing by default.
Pre-application	Applicants, utilities, governments	Basin-sensitive water statement, alternatives analysis, source hierarchy, early consultation.	Clarify assumptions before negotiations harden.
Permitting	Planning and environmental regulators	Caps, source restrictions, seasonal reporting, audit clauses, drought triggers, justice conditions.	Translate approval into enforceable obligations.
Construction and commissioning	Operators and utilities	Metering verification, infrastructure readiness review, public baseline publication.	Ensure the operating system matches approved claims.
Operation	Operators, regulators, utilities	Seasonal reporting, public dashboards, incident disclosure, independent audits.	Maintain accountability during routine use.

continued on following page

Table 7. Continued

Project stage	Lead actors	Illustrative instruments	Primary purpose
Stress and drought periods	Utilities, emergency managers, regulators	Curtailment rules, shortage communications, temporary restrictions, review hearings.	Protect fairness and legitimacy under crisis conditions.
Periodic review	All parties	Permit renewal tests, cumulative-impact reassessment, community oversight meetings.	Adapt governance as workloads and basin conditions change.

Figure 5. Actor responsibilities across the AI data center project life cycle.

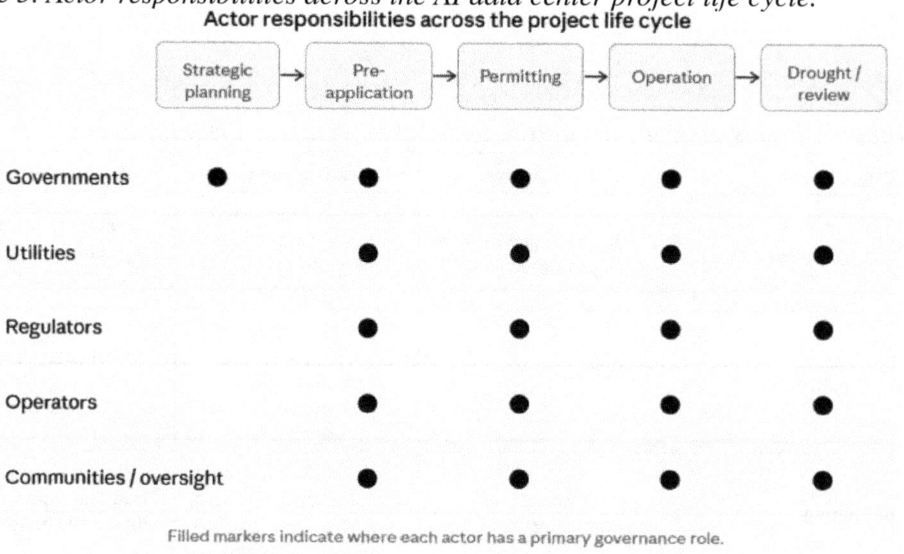

10. FUTURE RESEARCH DIRECTIONS

Several research priorities follow from the framework. First, more work is needed on measurement systems that connect facility performance to basin outcomes. Existing metrics are strongest at the level of withdrawals, consumption, or electricity use, but governance often requires coupled indicators that reveal how a facility behaves under seasonal stress, clustered development, or drought-stage escalation. Second, disclosure research should move beyond whether firms disclose water information toward whether specific disclosure architectures improve decision quality, public trust, and regulatory enforcement.

Third, comparative governance research is needed across jurisdictions. Water-stressed regions differ in utility governance, permitting authority, groundwater regulation, public-records law, and participation rights. Comparative analysis could clarify which institutional combinations are most effective in aligning digital-infrastructure growth with basin resilience. Fourth, future studies should integrate water, electricity, land, and justice indicators more systematically. Siloed metrics can make trade-offs look cleaner than they are. Fifth, more public facility-level datasets are required. Standardized, machine-readable data on withdrawals, consumption, source type, cooling systems, electricity procurement, and drought curtailment would substantially enhance both scholarship and civic oversight.

A sixth research priority concerns implementation capacity and administrative burden. Even when a jurisdiction adopts strong legal principles, water-sensitive governance can fail because agencies lack metering access, audit resources, hydrological expertise, or coordination routines across planning, utility, and emergency-management functions. More empirical work is needed on the institutional costs of high-quality disclosure, on the staffing models required for adaptive permit review, and on how smaller municipalities can govern digital infrastructure without being outmatched by technically sophisticated applicants. This is particularly important in regions where local governments compete for development but do not share basin-level analytical resources.

A seventh priority concerns the politics of review and enforcement. Future studies should examine how operators respond to reopener clauses, drought-triggered curtailment, public dashboards, and community oversight bodies in practice. It is also important to understand whether adaptive review improves trust or instead becomes vulnerable to political rollback once construction is complete. These questions point toward a broader agenda on institutional durability: the success of AI data center water governance will depend not only on well-designed frameworks, but on whether those frameworks remain enforceable under conditions of rapid market growth, changing drought severity, and uneven state capacity.

An eighth research priority concerns evaluation of policy sequencing. Not all instruments are equally valuable at every stage, and future scholarship should examine when exclusion zones, disclosure mandates, drought rules, community agreements, and adaptive permit review generate the greatest marginal governance benefits. Such work would help distinguish essential institutional capacities from desirable but secondary enhancements. It would also support more realistic guidance for jurisdictions with limited administrative resources, which may require phased governance strategies rather than attempting to implement every instrument simultaneously.

11. CONCLUSION

AI data centers are often described as an inevitable layer of modern economic development. This chapter argues that inevitability is the wrong governance frame, especially in water-stressed regions. The decisive issue is not whether digital infrastructure is valuable in general, but whether a specific project can operate within basin limits, under transparent accountability, and without generating unfair distributions of risk and voice. Framing the issue in this way shifts attention from abstract sustainability narratives to place-based public governance.

The revised chapter has advanced three connected commitments. Basin-aware siting ensures that approvals respond to cumulative hydrological conditions rather than parcel-level convenience. Transparency by design makes direct water use, indirect water dependence, drought behavior, and public commitments visible in forms that support comparison and enforcement. Water justice ensures that participation, recognition, burden sharing, and remedy are treated as core elements of legitimacy rather than symbolic supplements. Taken together, these commitments offer a more rigorous framework for evaluating when AI infrastructure should be approved, conditioned, resized, phased, or denied.

The broader implication is that responsible AI infrastructure is not only a technical accomplishment. It is also an institutional accomplishment. Regions that seek to host data centers in a sustainable way require planning systems capable of hydrological screening, utilities willing to participate in transparent scarcity governance, regulators able to impose adaptive conditions, and operators prepared to treat public accountability as part of routine operations. Under those conditions, digital development can be evaluated seriously and governed fairly. Without them, acceleration is not a development strategy on its own; it is a governance risk.

A final lesson is that governance quality should be assessed before controversy, not only after it. The earlier a jurisdiction clarifies hydrological limits, disclosure duties, and justice safeguards, the less likely it is that digital infrastructure approvals will be driven by opacity, urgency, or avoidable conflict.

REFERENCES

Ben-Amar, W., & Chelli, M. (2018). What drives voluntary corporate water disclosures? The effect of country-level institutions. *Business Strategy and the Environment, 27*(8), 1609–1622. DOI: 10.1002/bse.2227

Blackhurst, M., Hendrickson, C., & Vidal, J. S. I. (2010). Direct and indirect water withdrawals for U.S. industrial sectors. *Environmental Science & Technology, 44*(6), 2126–2131. DOI: 10.1021/es903147k PMID: 20141104

Bolón-Canedo, V., Morán-Fernández, L., Cancela, B., & Alonso-Betanzos, A. (2024). A review of green artificial intelligence: Towards a more sustainable future. *Neurocomputing, 565*, 128096. DOI: 10.1016/j.neucom.2024.128096

Burritt, R. L., Christ, K. L., & Omori, A. (2016). Drivers of corporate water-related disclosure: Evidence from Japan. *Journal of Cleaner Production, 129*, 65–74. DOI: 10.1016/j.jclepro.2016.04.119

Cannon, C. E. B., Chu, E. K., Natekal, A., & Waaland, G. (2024). Institutional designs for procedural justice and inclusion in urban climate change adaptation. *Journal of Planning Education and Research, 46*(1), 1–18. DOI: 10.1177/0739456X241274579

Farfan, J., & Lohrmann, A. (2023). Gone with the clouds: Estimating the electricity and water footprint of digital data services in Europe. *Energy Conversion and Management, 290*, 117225. DOI: 10.1016/j.enconman.2023.117225

Farooq, M. B., Naveed, K., Khalid, F., Narayan, A. K., & Khudir, I. M. (2025). Examining the extent and quality of corporate water management disclosures in extremely high-water stress countries. Sustainability Accounting. *Management and Policy Journal, 16*(3), 705–733. DOI: 10.1108/SAMPJ-01-2024-0054

Floridi, L., Cowls, J., Beltrametti, M., Chatila, R., Chazerand, P., Dignum, V., Luetge, C., Madelin, R., Pagallo, U., Rossi, F., Schafer, B., Valcke, P., & Vayena, E. (2018). AI4People—An ethical framework for a good AI society: Opportunities, risks, principles, and recommendations. *Minds and Machines, 28*(4), 689–707. DOI: 10.1007/s11023-018-9482-5 PMID: 30930541

Hoover, F.-A., Meerow, S., Grabowski, Z. J., & McPhearson, T. (2021). Environmental justice implications of siting criteria in urban green infrastructure planning. *Journal of Environmental Policy and Planning, 23*(5), 665–682. DOI: 10.1080/1523908X.2021.1945916

International Energy Agency. (2025). *Energy and AI*. Author.

Jin, Y., Behrens, P., Tukker, A., & Scherer, L. (2019). Water use of electricity technologies: A global meta-analysis. *Renewable & Sustainable Energy Reviews, 115*, 109391. DOI: 10.1016/j.rser.2019.109391

Jobin, A., Ienca, M., & Vayena, E. (2019). The global landscape of AI ethics guidelines. *Nature Machine Intelligence, 1*(9), 389–399. DOI: 10.1038/s42256-019-0088-2

Kuzma, S., Bierkens, M. F. P., Lakshman, S., Luo, T., Saccoccia, L., Sutanudjaja, E. H., & Van Beek, R. (2023). *Aqueduct 4.0: Updated decision-relevant global water risk indicators*. World Resources Institute.

Lannelongue, L., Grealey, J., & Inouye, M. (2021). Green algorithms: Quantifying the carbon footprint of computation. *Advanced Science (Weinheim, Baden-Württemberg, Germany), 8*(12), 2100707. DOI: 10.1002/advs.202100707 PMID: 34194954

Lee, U., Han, J., Elgowainy, A., & Wang, M. (2018). Regional water consumption for hydro and thermal electricity generation in the United States. *Applied Energy, 210*, 661–672. DOI: 10.1016/j.apenergy.2017.05.025

Li, P., Yang, J., Islam, M. A., & Ren, S. (2025). Making AI less "thirsty". *Communications of the ACM, 68*(7), 32–35. DOI: 10.1145/3724499

Macknick, J., Newmark, R., Heath, G., & Hallett, K. C. (2012). Operational water consumption and withdrawal factors for electricity generating technologies: A review of existing literature. *Environmental Research Letters, 7*(4), 045802. DOI: 10.1088/1748-9326/7/4/045802

Marantz, N. J., & Ulibarri, N. (2022). The tensions of transparency in urban and environmental planning. *Journal of Planning Education and Research, 42*(3), 401–412. DOI: 10.1177/0739456X19827638

Masanet, E., Shehabi, A., Lei, N., Smith, S., & Koomey, J. (2020). Recalibrating global data center energy-use estimates. *Science, 367*(6481), 984–986. DOI: 10.1126/science.aba3758 PMID: 32108103

McKee, E. (2025). Procedural justice in water management: A review. *Water (Basel), 17*(13), 1987. DOI: 10.3390/w17131987

Mittelstadt, B. (2019). Principles alone cannot guarantee ethical AI. *Nature Machine Intelligence, 1*(11), 501–507. DOI: 10.1038/s42256-019-0114-4

Mytton, D. (2021). Data centre water consumption. npj. *NPJ Clean Water, 4*(1), 11. DOI: 10.1038/s41545-021-00101-w

Pinheiro, A. B., do Prado, N. B., & de Moraes, G. H. S. M. (2024). What drives corporate water disclosure? The role of board composition in Brazil and India. *Sustainable Environment, 10*(1), 2426836. DOI: 10.1080/27658511.2024.2426836

Shahrad, A., Stålhammar, S., & Gulsrud, N. M. (2025). Procedural justice for whom? A case study of a densification project. *Local Environment, 30*(2), 546–566. DOI: 10.1080/13549839.2024.2447757

Shrimpton, E. A., Hunt, D. V. L., & Rogers, C. D. F. (2021). Justice in (English) water infrastructure: A systematic review. *Sustainability (Basel), 13*(6), 3363. DOI: 10.3390/su13063363

Stillwell, A. S., & Webber, M. E. (2014). Geographic, technologic, and economic analysis of using reclaimed water for thermoelectric power plant cooling. *Environmental Science & Technology, 48*(19), 11160–11168. DOI: 10.1021/es405820j PMID: 24625241

Strubell, E., Ganesh, A., & McCallum, A. (2019). Energy and policy considerations for deep learning in NLP. In *Proceedings of the 57th Annual Meeting of the Association for Computational Linguistics* (pp. 3645-3650). DOI: 10.18653/v1/P19-1355

Talati, S., Zhai, H., Kyle, G. P., Morgan, M. G., Patel, P., & Liu, L. (2016). Consumptive water use from electricity generation in the Southwest under alternative climate, technology, and policy futures. *Environmental Science & Technology, 50*(16), 9019–9027. DOI: 10.1021/acs.est.6b01389 PMID: 27768843

Verdecchia, R., Sallou, J., & Cruz, L. (2023). A systematic review of green AI. *Wiley Interdisciplinary Reviews. Data Mining and Knowledge Discovery, 13*(4), e1507. DOI: 10.1002/widm.1507

Wahyuningrum, I. F. S., Budihardjo, M. A., Rustianto, A., & Handayani, N. U. (2023). Determinants of corporate water disclosure in Indonesia. *Sustainability (Basel), 15*(14), 11107. DOI: 10.3390/su151411107

Wang, J., Ulibarri, N., Scott, T. A., & Davis, S. J. (2023). Environmental justice, infrastructure provisioning, and environmental impact assessment: Evidence from the California Environmental Quality Act. *Environmental Science & Policy, 146*, 66–75. DOI: 10.1016/j.envsci.2023.05.003

Yu, H.-C., Kuo, L., & Ma, B. (2020). The drivers of corporate water disclosure in enhancing information transparency. *Sustainability (Basel), 12*(1), 385. DOI: 10.3390/su12010385

Zhang, L., Tang, Q., & Huang, R. H. (2021). Mind the gap: Is water disclosure a missing component of corporate social responsibility? *The British Accounting Review, 53*(1), 100940. DOI: 10.1016/j.bar.2020.100940

ADDITIONAL READING

International Energy Agency. (2025). *Data centres and data transmission networks*. Author.

Kuzma, S., Saccoccia, L., & Chertock, M. (2023). *25 countries, housing one-quarter of the population, face extremely high water stress*. World Resources Institute.

Li, P., Yang, J., Islam, M. A., & Ren, S. (2025). Making AI less "thirsty". *Communications of the ACM, 68*(7), 32–35. DOI: 10.1145/3724499

Marantz, N. J., & Ulibarri, N. (2022). The tensions of transparency in urban and environmental planning. *Journal of Planning Education and Research, 42*(3), 401–412. DOI: 10.1177/0739456X19827638

Mytton, D. (2021). Data centre water consumption. npj. *NPJ Clean Water, 4*(1), 11. DOI: 10.1038/s41545-021-00101-w

UN-Water. (2024). *United Nations world water development report 2024: Water for prosperity and peace*. UNESCO.

KEY TERMS AND DEFINITIONS

Adaptive permitting: A permit structure that includes predefined review triggers linked to changes in basin conditions, operating assumptions, electricity sourcing, or cumulative regional demand. Adaptive permits allow public authorities to revisit and revise conditions without requiring a full new application, so that governance remains aligned with evolving hydrological and social realities.

AI data center: A facility or campus providing large-scale compute, storage, and networking capacity for AI training, inference, and related digital services. Distinguished from conventional data centers by high power density, continuous high-availability operation, and cooling demands that scale with AI workloads.

Basin-aware siting: A site-selection approach that evaluates the hydrological, social, and infrastructural context of the receiving basin before project approval is granted. Basin-aware siting goes beyond parcel-level feasibility to assess cumulative water demand, drought exposure, source suitability, and community vulnerability at the watershed scale.

Caution zone: A basin or service area designated as available for potential development but requiring enhanced review, stronger proof of hydrological compatibility, tighter operating constraints, and more demanding public oversight before approval. Distinct from an exclusion zone, where development is presumptively unsuitable.

Cumulative basin demand: The aggregate water withdrawal and consumption resulting from the interaction of all current and projected users within the same basin or service area. Cumulative demand is the relevant unit for assessing cluster effects and long-run allocation pressure, and cannot be approximated solely from parcel-level data.

Direct water use: Water withdrawn, recycled, discharged, or consumed on site for cooling, humidification, and related facility operations. Direct water use is measured through metered facility withdrawals, seasonal consumption profiles, and source-specific reporting, and serves as the primary permit-control variable in basin-sensitive governance.

Distributive justice: A dimension of water justice concerned with how benefits and burdens are allocated across different groups. In the context of AI data center governance, distributive justice asks who receives durable benefits such as jobs and tax revenues, and who absorbs water-system risks such as supply competition, affordability pressure, and drought burden.

Drought-stage performance: The documented behavior of a facility under heat extremes, shortage declarations, or emergency curtailment conditions. Used as an adaptive permit trigger and public accountability metric to determine whether a facility can coexist with basin-level scarcity management.

Exclusion zone: A basin or service area designated as presumptively unsuitable for water-intensive development due to chronic scarcity, heavy groundwater overdraft, ecological fragility, or acute inequality in water service. Distinct from a caution zone, where development may be permitted under enhanced review.

Indirect water use: Water embodied in electricity generation or other upstream services required to operate a facility. Indirect water use is disclosed as a contextual and planning indicator rather than a direct permit-control variable, unless a jurisdiction has the legal authority to condition electricity sourcing directly.

Power usage effectiveness (PUE): A common data center efficiency metric that expresses total facility energy consumption as a ratio to IT equipment energy consumption. A PUE of 1.0 indicates that all energy goes directly to computing; higher values indicate proportionally greater overhead consumption. PUE captures energy overhead but does not reflect water use, basin conditions, or social impact.

Procedural justice: A dimension of water justice concerned with the fairness of decision-making processes. Procedural justice requires early community engagement, intelligible information, meaningful opportunities to influence key terms of approval before those terms become path dependent, and institutional support for participation by less-resourced stakeholders.

Reclaimed water (recycled water): Treated wastewater or other non-potable water sources used in place of freshwater for cooling or other industrial purposes. Reclaimed water is often presented as a sustainability solution for data center cooling, but its governance value depends on infrastructure reliability, competing uses, and seasonal availability rather than its designation alone.

Recognitional justice: A dimension of water justice concerned with whether the identities, histories, and knowledge of affected communities are acknowledged and respected in governance processes. Recognitional justice requires taking local experience seriously, providing language access, and acknowledging that formally equal participation may still reproduce inequality when actors differ in resources and institutional access.

Remedial justice: A dimension of water justice concerned with what happens when commitments, performance standards, or community agreements are not met. Remedial justice requires that permits and agreements specify audits, penalties, permit reopeners, and revisable approvals rather than relying on aspirational pledges without enforcement mechanisms.

Social license: The ongoing acceptance and legitimacy granted by affected communities and stakeholders to a project or operator, based on demonstrated adherence to procedural, distributive, recognitional, and remedial standards. Social license is not a one-time approval but a sustained relationship between operators and the communities that host their infrastructure.

Source hierarchy: A ranked ordering of water supply sources (e.g., reclaimed water, groundwater, surface water, potable water) that a facility is authorized to use, specifying which sources are primary, which are backup, and under what conditions each may be accessed. Source hierarchies are set in permits to prioritize less scarce or less socially sensitive supplies.

Transparency by design: A governance approach in which disclosure obligations are built into approvals and operations from the start rather than added after controversy emerges. Transparency by design requires application-stage disclosure, periodic public reporting, prompt incident notices, and public dashboards that allow regulators and communities to evaluate performance over time.

Water consumption: The portion of withdrawn water that is not returned to the source watershed, typically because it has evaporated, been incorporated into a product, or been discharged to a different basin. Water consumption is a more complete measure of basin-level impact than withdrawal alone, and is the preferred metric for assessing facility contributions to water scarcity.

Water justice: A governance approach that evaluates participation, burden sharing, recognition, and remedy in water-related decisions. In this chapter, water justice encompasses four dimensions—procedural, distributive, recognitional, and remedial—and is treated as a core element of legitimacy rather than a supplementary concern.

Water stress: A condition in which scarcity, variability, ecological limits, or allocation pressure make additional withdrawals from a basin socially or environmentally sensitive. Water stress is highly place specific and should be assessed through local and regional analysis rather than solely through global or portfolio-level indicators.

Water use effectiveness (WUE): An industry metric that expresses a data center's on-site water consumption (in liters) per unit of IT energy delivered (in kilowatt-hours). WUE is useful for comparing facility-level efficiency but does not capture indirect water dependence, source type, basin conditions, seasonal variation, or drought behavior, and therefore requires supplementation with broader governance indicators.

Water withdrawal: The total volume of water extracted from a source (surface water, groundwater, or utility supply) for use in facility operations, regardless of how much is subsequently returned. Withdrawal figures indicate the scale of a facility's demand on the supply system and are typically the primary metric reported in operating permits.

Chapter 8
Water and Energy Sustainability in Traditional and AI-Driven Data Centers:
Environmental, Economic, and Policy Implications

Usman Abidemi Sarumi
http://orcid.org/0009-0003-0784-8055
AI and Software Engineering Dept., AI and IoT Research Center, AI and Informatics Faculty, Near East University, Mersin, Turkey

Muhammed Murat Yanaşoğlu
http://orcid.org/0009-0007-4722-8079
AI and Software Engineering Dept., AI and IoT Research Center, AI and Informatics Faculty, Near East University, Mersin, Turkey

Cliford Peter Mlimwa
http://orcid.org/0009-0000-3071-8847
AI and Software Engineering Dept., AI and IoT Research Center, AI and Informatics Faculty, Near East University, Mersin, Turkey

Sinem Alturjman
AI and Software Engineering Dept., AI and IoT Research Center, AI and Informatics Faculty, Near East University, Mersin, Turkey

Fadi Al-Turjman
http://orcid.org/0000-0001-6375-4123
AI and Software Engineering Department, AI and IoT Research Center, AI and Informatics Faculty, Near East University, Mersin, Turkey

Hitesh Mohapatra
http://orcid.org/0000-0001-8100-4860
School of Computer Engineering, Kalinga Institute of Industrial Technology (Deemed to be University), Bhubaneswar, India

DOI: 10.4018/979-8-2600-2761-5.ch008

ABSTRACT

This chapter analyze the environmental, economic, and policy integration of water and energy sustainability in conventional and AI-optimized data centers. The authors seek to study the impacts of architectural decisions and AI optimization on the sustainability performance of data centers focusing on the power usage effectiveness and water-use intensity as the main sustainability metrics and the sustainability trade-offs in the structures capital cost, operational cost, and total cost of ownership over time. The chapter reviews available industry and academic works on the cooling and workload management and driving factors on the increasing water and energy consumption of AI workloads and offers a policy and governance framework to improve the speed of the efficiencies. The responsible use of AI in data centers has the potential to lower operational water and energy consumption. However, AI in data centers has the potential to displace the impacts of sustainability, leading to the justification of the need for the optimization of sustainability processes.

INTRODUCTION

Background

With the incorporation of cloud computing, AI, big data, and IoT, data centers have become core infrastructure for contemporary digital technologies. The global data center market has increased rapidly, growing from approximately 500,000 sites and facilities in 2015 to over 8 million in 2024 (Guidi et al., 2024). The digitization of industries, together with the growth of AI workloads, has created demand for instantaneous data processing and storage, placing pressure on data center energy, cooling, and water capacity. Therefore, data center sustainability must consider not only energy efficiency but also water use, carbon emissions, economic performance, and policy readiness.

Data centers that rely on traditional cooling systems and static resource allocation typically report Power Usage Effectiveness (PUE) values of 1.4 to 1.8. PUE measures the ratio of total facility energy to the energy used directly by IT equipment; therefore, a PUE of 1.4 to 1.8 means that for every watt consumed by IT equipment, an additional 0.4 to 0.8 watts are used for cooling, power distribution, and other supporting systems (Kannan et al., 2025). Data centers that incorporate AI use machine learning, predictive analytics, and intelligent automation to balance energy, cooling, and resource demand (Qi et al., 2023b; Qi et al., 2023a; Tuli et al., 2021). However, AI model training and inference can create a sustainability paradox in which data centers become more operationally efficient while also becoming more computationally intensive (Singh et al., 2024; Jegham et al., 2025; Amati et al., 2024).

Problem Statement

Data centers are confronted with multifaceted sustainability challenges that are +operational, environmental, and regulatory. Many traditional architectures contain structural inefficiencies because infrastructure is sized for peak-load operation even when facilities use only 10-30% of their available operational capacity. In regions where water scarcity and carbon-emission concerns are significant, these inefficiencies increase pressure on operators and intensify regulatory scrutiny (Li et al., 2023a; Guidi et al., 2024; Han et al., 2024).

The economic practicality of operating older facilities with high PUE values continues to decline as energy and water costs increase (Hogade & Pasricha, 2024). At the same time, many policy landscapes remain incomplete or underdeveloped, meaning that regulators often provide insufficient or inconsistent guidance on water-consumption reporting, greenhouse gas emissions, and the sustainability attributes of AI-enabled data centers (Luccioni et al., 2025; Ebert et al., 2024; Mabrouk et al., 2025).

Research Objectives

This chapter presents a comparative analysis of traditional and AI-centered data centers. It uses the three research questions below to compare both models within a sustainability framework that considers energy, water, carbon, economic performance, and policy alignment. It also weighs AI optimization against trade-offs such as computational overhead, water consumption, and rebound effects.

RQ1: How do traditional and AI-based data center models compare in terms of energy use, water consumption, and CO2 emissions under similar operating conditions? The analysis relies on PUE, WUE, and CUE. WUE refers to water consumed per unit of IT energy, while CUE refers to carbon emissions per unit of IT energy (Qi et al., 2023b; Qi et al., 2023a; Jiang et al., 2025c; Jiang et al., 2025b; Naug et al., 2025; Kannan et al., 2025; Becchetti et al., 2025).
RQ2: How do traditional and AI-based data center models compare in terms of capital investment, TCO, ROI, and payback period? TCO is the sum of capital and operating costs across the useful life of the asset. ROI is the financial return from efficiency improvements, and the payback period is the time required for savings to equal the initial investment (Hogade & Pasricha, 2024; Li et al., 2025; Bahi & Ourici, 2025; Amati et al., 2024).

RQ3: Where do current policies align or diverge in the data center sustainability ecosystem, and how can policy frameworks balance environmental protection with economic competitiveness? This question addresses sustainability reporting, resource caps, and innovation incentives in data center sustainability policy (Ebert et al., 2024).

Scope and Limitations

This study focuses on large enterprise and hyperscale data centers with IT loads greater than 1 MW. It compares traditional system architectures with AI/ML-based architectures. The geographic scope is North America and Europe, and the period of interest is 2021 to 2025.

The limitations of this study are as follows. First, the analysis relies on publicly available data and benchmark estimates rather than proprietary operational data (Guidi et al., 2024; Kannan et al., 2025). Second, rapid technological advancement may affect the relevance of some findings over time. Third, the economic analysis is based on present-day market assumptions that may change (Hogade & Pasricha, 2024; Li et al., 2025). Fourth, the analysis focuses on operational effects rather than a full life cycle assessment (LCA), which would include construction, operation, and decommissioning impacts. Fifth, policy analysis is limited by regional regulatory variation (Ebert et al., 2024). Finally, many institutions use hybrid architectures and are therefore not completely traditional or fully AI-based (Tuli et al., 2021; Panteleaki et al., 2025; Sarkar et al., 2025; Schmitz et al., 2025; Becchetti et al., 2025). Despite these limitations, the study offers useful contributions to sustainable data center design and policy-gap analysis.

LITERATURE REVIEW

Conventional Data Center Energy and Water Use

The typical Power Usage Effectiveness (PUE) of traditional data centers falls between 1.4 and 1.8. Based on the literature, approximately 40-50% of data center energy consumption is used by IT equipment, 30-40% by cooling systems, and 10-20% by power distribution (Kannan et al., 2025; Ajibola et al., 2021; Saxena & Singh, 2022). Greater reliance on air-based and evaporative cooling solutions can increase water consumption in water-scarce areas (Li et al., 2023a). In addition, the carbon impact of data centers depends on local electricity generation and consumption patterns (Guidi et al., 2024; Han et al., 2024; Li et al., 2023b). Many legacy systems allocate power and cooling resources as if systems operate at full capacity, even though many use only 10-30% of available resources. This creates unnecessary and excessive end-to-end resource consumption.

AI-Driven Data Center Optimization

AI and machine learning are applied to advanced methods such as predictive cooling, workload forecasting, dynamic resource allocation, and multi-objective optimization (Qi et al., 2023b; Qi et al., 2023a; Tuli et al., 2021). When combined with effective resource management, these forecasting and cooling-control approaches can achieve PUE values of 1.1-1.3 (Naug et al., 2025; Kannan et al., 2025). The SHIELD and MOSAIC frameworks demonstrate the simultaneous optimization of carbon, wastewater, and energy objectives (Qi et al., 2023b; Qi et al., 2023a). Energy cost and carbon intensity can also be mitigated in distributed settings through reinforcement learning and workload placement (Hogade & Pasricha, 2024). Liquid-cooling optimization extends these methods to data center thermal control (Sarkar et al., 2025; Naug et al., 2025). Nonetheless, sustainability assessment must account for the substantial energy and water overhead of AI workloads, especially large language model training and inference (Li et al., 2023a; Jiang et al., 2025b; Singh et al., 2024; Jegham et al., 2025; Mitra et al., 2023).

Sustainability Metrics and Multi-Objective Assessment

Power Usage Effectiveness (PUE) remains the most widely used metric for evaluating data center energy efficiency because it relates total facility energy consumption to the energy used directly by IT equipment. Lower PUE values indicate that less energy is consumed by cooling, power distribution, and other overhead systems. AI-based prediction and cooling-load models use this metric to evaluate improvements from predictive control (Kannan et al., 2025; Jiang et al., 2025a).

Water Usage Effectiveness (WUE) and Carbon Usage Effectiveness (CUE) are increasingly important because energy-efficient facilities can still create water stress or carbon-intensive impacts. WUE measures water consumed per unit of IT energy, while CUE measures carbon emissions per unit of IT energy (Qi et al., 2023b; Qi et al., 2023a; Hogade & Pasricha, 2024). WaterWise and ThirstyFLOPS illustrate the growing focus on water-carbon co-optimization and water footprint modeling in cloud and high-performance computing environments (Li et al., 2023a; Jiang et al., 2025c; Jiang et al., 2025b). Other studies show that carbon-aware workload redistribution, renewable integration, and geographic load balancing can reduce emissions when supported by appropriate energy-management strategies (Hogade & Pasricha, 2024; Li et al., 2023b; Li et al., 2025; Bahi & Ourici, 2025), while newer metric architectures integrate performance, efficiency, and cost across AI infrastructure layers (He, 2025).

Research Gaps and Positioning of This Chapter

The reviewed literature reveals six gaps that motivate the present chapter: (1) limited comparative studies evaluating traditional and AI-driven data centers across energy, water, and carbon within the same framework (Qi et al., 2023b; Qi et al., 2023a; Tuli et al., 2021); (2) insufficient attention to the water footprint of AI-enabled training and inference workloads (Li et al., 2023a; Jiang et al., 2025b; Jegham et al., 2025); (3) limited integration of TCO, ROI, and payback analysis with environmental assessment (Hogade & Pasricha, 2024; Li et al., 2025; Bahi & Ourici, 2025); (4) weak policy guidance for AI-specific facility characteristics and sustainability reporting (Luccioni et al., 2025; Ebert et al., 2024); (5) under-examined rebound effects and Jevons' paradox in AI efficiency gains; and (6) limited lifecycle assessment of embodied energy, water use, construction, manufacturing, and decommissioning. Table 1 summarizes the reviewed studies according to focus area, key findings, and metrics used.

Table 1. Literature Review Summary by Theme

Study	Focus Area	Key Findings	Metrics Used
Qi et al. (2023)	Multi-objective	SHIELD framework achieves simultaneous optimization of	PUE, CUE, WUE
Qi et al. (2023)	Sustainable data center	MOSAIC framework demonstrates multi-objective optimization	PUE, CUE, WUE
Li et al. (2023)	Water footprint of AI	AI model training consumes millions of gallons of water;	WUE, Water
Tuli et al. (2021)	AI-based resource	HUNTER framework provides holistic resource management for	PUE, Energy
Jiang et al. (2025)	Water–carbon co-	WaterWise framework co-optimizes carbon and water footprints	WUE, CUE
Jiang et al. (2025)	Water footprint modeling	ThirstyFLOPS provides comprehensive water footprint analysis	WUE, Water
Guidi et al. (2024)	Environmental burden	US data centers contribute 100–150 million metric tons CO_2	CUE, Carbon

continued on following page

Table 1. Continued

Study	Focus Area	Key Findings	Metrics Used
Hogade & Pasricha (2024)	Carbon and energy	Game-theoretic deep RL minimizes carbon emissions and energy	CUE, Energy costs
Han et al. (2024)	Public health impact	Quantifies public health impacts of AI data center	Carbon emissions
Li et al. (2023)	Geographic load balancing	Environmentally equitable AI through geographic workload	CUE, Carbon
Singh et al. (2024)	LLM sustainability	Survey of sustainability challenges in large language models	Energy, Water, Carbon
Luccioni et al. (2025)	Rebound effects	Examines Jevons' paradox in AI efficiency improvements	Efficiency metrics
Ebert et al. (2024)	Policy and regulation	Analyzes policy gaps in data center sustainability	Policy frameworks
Panteleaki et al. (2025)	Edge data centers	Vertical approach to sustainable heterogeneous edge data	PUE, Energy
Sarkar et al. (2025)	Liquid cooling	Hierarchical multi-agent framework for carbon-efficient	CUE, PUE
Li et al. (2025)	Renewable energy	Energy management for renewable-colocated AI data centers	Renewable energy
Bahi & Ourici (2025)	Green energy integration	Deep RL for real-time green energy integration	Renewable energy
Jegham et al. (2025)	LLM inference footprint	Benchmarks energy, water, and carbon footprint of LLM	Energy, Water, Carbon
Naug et al. (2025)	Liquid cooling	LC-Opt benchmarks RL for end-to-end liquid cooling	PUE, Cooling
Kannan et al. (2025)	PUE prediction	Bidirectional GRU model for accurate PUE prediction	PUE
Zhan et al. (2025)	Cooling optimization	Offline RL for data center cooling system optimization	PUE, Cooling
Jiang et al. (2025)	Cooling load prediction	HyperLoad uses LLM for green data center cooling load	Cooling load, PUE

continued on following page

Table 1. Continued

Study	Focus Area	Key Findings	Metrics Used
Liu et al. (2025)	Federated learning	Adaptive federated learning optimizes integrated flows in	Energy efficiency
Ajibola et al. (2021)	Workload placement	Energy-efficient workload placement in composable data	PUE, Energy
Saxena & Singh (2022)	VM allocation	Proactive autoscaling and energy-efficient VM allocation	PUE, Energy
Quach et al. (2025)	GPU co-location	ECLIP: energy-efficient practical co-location of ML	Energy, Throughput
Liu & Mauter (2021)	Water supply energy	Marginal energy intensity of water supply	Energy intensity
Silva-Rodriguez & Li (2023)	Water–energy co-	Decentralized micro water-energy co-optimization for small	Energy, Water
Goodarzi & Li (2022)	Energy–water nexus	Security-constrained optimal operation of energy–water nexus	Cost, Reliability
Obringer et al. (2023)	Climate & water–energy	Climate analogs project polarization of urban water–energy	Water–energy
Alahmed et al. (2025)	Power–water co-scheduling	Co-scheduling power and water in desalination plants	Energy cost, Water
Schmitz et al. (2025)	Energy security &	Reviews energy security/ resilience concepts and planning	Resilience metrics
Mitra et al. (2023)	Hydropower modeling	Gaps in representations of hydropower generation in models	Modeling metrics
Saheb & Dehghani (2021)	AI for sustainable energy	Contextual topic modeling and content analysis	Bibliometrics
Di Martino et al. (2023)	Energy–water systems	Optimal energy–water supply system operation for	Cost, Energy
Mabrouk et al. (2025)	Multi-carrier hubs	Optimal scheduling of energy and mass flows in networked	Cost, Flow
He (2025)	AI infrastructure metrics	Cross-layer taxonomy integrating performance, efficiency,	Performance

continued on following page

Table 1. Continued

Study	Focus Area	Key Findings	Metrics Used
Becchetti et al. (2025)	PV & storage investment	Dynamic model to advance investments in PV and energy	ROI/NPV
Amati et al. (2024)	Energy efficiency modeling	Energy efficiency study using Lattice Boltzmann method	Efficiency metrics
Hasanbeig et al. (2022)	Certified RL	LCRL: Certified policy synthesis via logically-constrained RL	Safety/Constraints

- PUE: Power Usage Effectiveness
- WUE: Water Usage Effectiveness
- CUE: Carbon Usage Effectiveness
- RL: Reinforcement Learning
- LLM: Large Language Model
- HPC: High-Performance Computing

METHODOLOGY

Comparative Framework

This research utilizes a comparative approach to analyze traditional and AI-enhanced data center models across environmental, economic, and policy dimensions. For comparison purposes, both models are analyzed at the same facility scale, with similar workloads and the same sustainability parameters. The approach links three dimensions: environmental performance, including energy, water, and carbon impacts; economic performance, including TCO, ROI, OPEX, and payback period; and policy readiness, including reporting, resource management, and regulatory gaps.

The assessment employs a multi-metric approach because improvement in one metric may lead to deterioration in another. A facility may improve Power Usage Effectiveness (PUE), but sustainability may still decline if water demand or total electricity demand increases through AI workloads. WUE, CUE, and PUE are therefore evaluated together rather than in isolation (Qi et al., 2023b; Qi et al., 2023a; Jiang et al., 2025c).

This research focuses on models under normal operating conditions. Data center models are assessed using the same IT-load capacity (>1 MW), similar geographic and climatic conditions, similar workloads, and the same assessment timeframe.

Data Sources

This analysis incorporates industry benchmarks, peer-reviewed literature, contemporary literature from 2021-2025, and relevant policy frameworks. The Uptime Institute, The Green Grid Consortium, and the International Energy Agency (IEA) provide industry resources based on standardized data center operational benchmarks (Guidi et al., 2024; Kannan et al., 2025). Academic and technical studies address AI-enabled optimization, liquid cooling, workload placement, water footprint modeling, and carbon-aware resource management (Qi et al., 2023b; Qi et al., 2023a; Tuli et al., 2021; Jiang et al., 2025b; Hogade & Pasricha, 2024; Han et al., 2024; Li et al., 2023b).

Published case studies, white papers, and policy analyses are the primary sources used to contextualize facility-scale metrics, economic assumptions, and regulatory constraints. In the absence of proprietary operational data, these sources are treated as benchmarked and literature-supported estimates rather than direct facility measurements. Policy sources inform the analysis of water and carbon reporting, as well as AI-specific sustainability governance frameworks (Ebert et al., 2024).

Metrics and Indicators

To evaluate the two data center models transparently, commonly accepted sustainability and economic indicators were adopted. Power Usage Effectiveness (PUE) compares the total energy consumed by the entire facility with the energy consumed by IT equipment. A lower PUE indicates that overhead systems, including infrastructure, energy distribution, and cooling systems, consume less energy (Kannan et al., 2025).

Water Usage Effectiveness (WUE) and Carbon Usage Effectiveness (CUE) are used to quantify resource efficiency and emissions intensity. WUE is expressed in liters per kWh, and CUE is expressed in kilograms of CO2 equivalent per kWh. These metrics must be used together because cooling design, workload distribution, and electricity sourcing can affect electricity consumption, water consumption, and carbon output (Li et al., 2023a; Jiang et al., 2025c; Jiang et al., 2025b; Qi et al., 2023b; Qi et al., 2023a; Hogade & Pasricha, 2024).

The financial analysis employs Total Cost of Ownership (TCO), Return on Investment (ROI), and the payback period. TCO refers to the initial investment and ongoing operational costs; ROI quantifies the financial benefits of efficiency-based investments; and the payback period determines the duration necessary for the achieved savings to offset (recover) the initial investment (Hogade & Pasricha, 2024; Li et al., 2025; Bahi & Ourici, 2025).

Table 2 defines the metrics, formulas, units, and target ranges used for comparison.

Analytical Procedure and Assumptions

The analysis is based on a series of steps. First, representative resource-management benchmarks and ranges are identified. Second, AI-based and non-AI-based models are separated according to their resource-management logic. Third, both models are evaluated using the same criteria. Fourth, differences are expressed as percentage improvements. The final steps identify AI-related trade-offs, including training, inference, monitoring, and updating, and interpret policies related to resource constraints, reporting, and incentives (Singh et al., 2024; Jegham et al., 2025; Hasanbeig et al., 2022; Qi et al., 2023b; Qi et al., 2023a).

An example can be shown with a 10 MW IT-load facility. A traditional design with a PUE of 1.6 requires 16 MW of total facility energy, while an AI-enabled design with a PUE of 1.2 requires 12 MW. This represents a 25% reduction in total facility energy. The same approach is used for WUE, CUE, TCO, ROI, and payback period while considering sensitivity to energy pricing, water availability, carbon pricing, and technological maturity (Hogade & Pasricha, 2024; Li et al., 2025; Kannan et al., 2025).

Table 2. Sustainability Metrics and Formulas

Metric	Definition	Formula	Unit	Target Range
PUE (Power Usage Effectiveness)	Ratio of total facility energy to IT equipment energy	PUE = Total Facility Energy / IT Equipment Energy	Dimensionless	Excellent: 1.1-1.2 Good: 1.2-1.4 Average: 1.4-1.6 Poor: >1.6
WUE (Water Usage Effectiveness)	Water consumption per unit of IT equipment energy	WUE = Total Water Consumption (L) / IT Equipment Energy (kWh)	L/kWh	Lower is better Target: <0.5 L/kWh
CUE (Carbon Usage Effectiveness)	Carbon emissions per unit of IT equipment energy	CUE = Total CO_2 Emissions (kg CO_2e) / IT Equipment Energy (kWh)	kg CO_2e/kWh	Lower is better Target: <0.5 kg CO_2e/kWh
TCO (Total Cost of Ownership)	Sum of all costs over facility lifetime	TCO = Initial Capital + Σ(Annual Operational Costs × Years)	Currency (USD)	Varies by facility size and location

continued on following page

Table 2. Continued

Metric	Definition	Formula	Unit	Target Range
ROI (Return on Investment)	Ratio of net benefits to initial investment	ROI = (Net Benefits / Initial Investment) × 100%	Percentage (%)	Positive ROI indicates economic viability
Payback Period	Time for cumulative savings to equal initial investment	Payback Period = Initial Investment / Annual Savings	Years	Shorter is better Target: <5 years

Calculation Assumptions:

- Evaluation Period: 10 years for TCO, ROI, and payback comparison.
- Energy Price: Variable by geographic location, grid mix, and market conditions.
- Water Tariff: Variable by regional water availability and local scarcity conditions.
- Carbon Pricing: Based on regional carbon markets and regulatory requirements.
- Discount Rate: 5% for present-value and long-term economic comparison.

TRADITIONAL DATA CENTERS: INFRASTRUCTURE AND RESOURCE CONSUMPTION

According to Tuli et al. (2021) and Hasanbeig et al. (2022), traditional data center architectures are designed with three infrastructure layers: data communication and computing equipment, cooling support infrastructure, and electrical power-distribution infrastructure. The data communication and computing layer includes storage and networking devices as well as server racks, which are typically located in raised-floor rooms with overhead or underfloor power and network distribution (Ajibola et al., 2021; Saxena & Singh, 2022).

In traditional facilities, evaporative cooling towers and air systems are commonly employed. Computer room air conditioning and air handling units circulate cooled air through the raised floor to help maintain room temperatures within the ASHRAE recommended range of 18-27 °C. Evaporative cooling towers account for a major part of facility water consumption because they use water to reject heat (Li et al., 2023a; Liu & Mauter, 2021).

In addition to UPS systems, PDUs, and backup generators, electrical power-distribution systems reduce energy efficiency because electricity undergoes several conversions. Typical power-distribution efficiency is between 85% and 92%, meaning that distribution systems can lose between 8% and 15% of energy as waste heat (Kannan et al., 2025; Ajibola et al., 2021).

Figure 1 shows the normal arrangement of a traditional data center, depicting how IT equipment, cooling systems, and power-distribution elements interact to determine facility energy and resource consumption.

Figure 1. Traditional Data Center Architecture and Resource-Flow Diagram

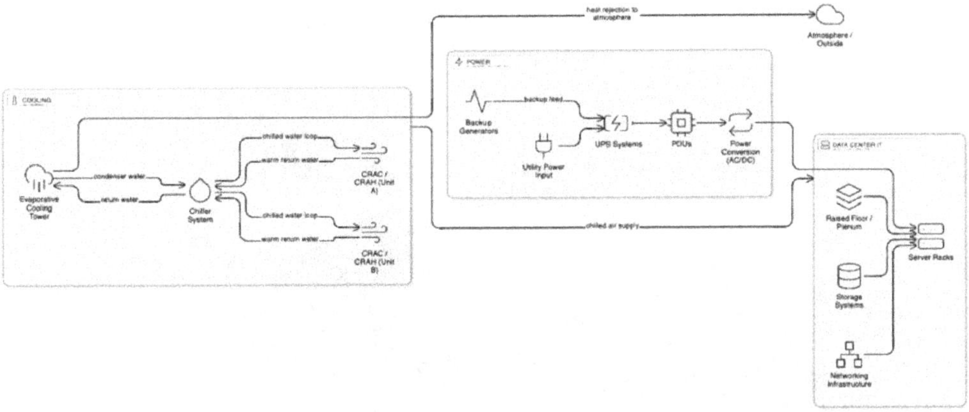

Energy Consumption Trends

In conventional data centers, IT equipment accounts for approximately 40%-50% of total facility energy consumption. Cooling systems account for approximately 30%-40%, and power-distribution systems account for the remaining 10%-20% (Kannan et al., 2025; Ajibola et al., 2021; Saxena & Singh, 2022).

Conventional data centers exhibit high levels of underutilization. Average server utilization is approximately 10%-30%, and servers continue to consume a significant amount of power even during idle periods (Tuli et al., 2021; Ajibola et al., 2021). Computing resources are often allocated statically, so capacity is provisioned for peak demand even when normal demand is lower. This creates resource waste during lower-demand operation, and idle servers may still draw substantial power (Saxena & Singh, 2022).

Cooling systems typically operate with limited adaptive control. Fixed set points are often conservative and may provide more cooling than necessary to maintain safe operating conditions.

Operational losses in power-distribution systems occur at the utility-to-UPS (5%-8%), UPS-to-PDU (2%-4%), and PDU-to-server (3%-5%) levels. These compounded losses create significant energy waste and increase the cooling burden (Kannan et al., 2025; Ajibola et al., 2021).

PUE values in the range of 1.4-1.8 imply that traditional data centers consume 0.4-0.8 additional watts for cooling, power distribution, and supporting infrastructure for every watt consumed by IT equipment (Kannan et al., 2025; Ajibola et al., 2021). Figure 2 illustrates the energy-consumption characteristics of typical data centers.

Figure 2. Traditional Data Center Energy Consumption Distribution

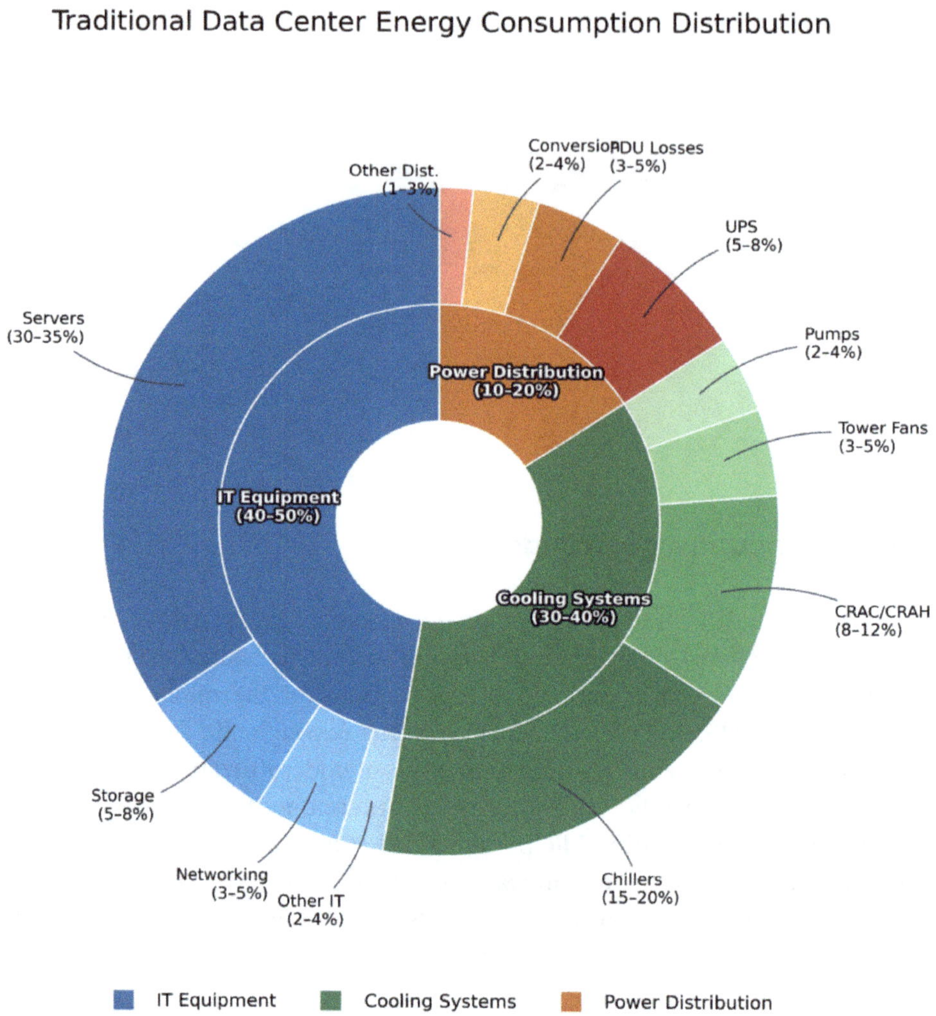

Analysis of Water Consumption

Water use in traditional data centers is largely associated with cooling systems. Large facilities with IT loads greater than 10 MW may use more than 1 million gallons of water per day for cooling.

Evaporative towers use water to cool heat exchangers, and a portion of that water evaporates to remove heat. The evaporated water must be replaced, creating a direct relationship between cooling load and water consumption (Li et al., 2023a; Liu & Mauter, 2021).

Periodic water replacement to prevent mineral build-up, known as blowdown, and water loss due to drift can add 5%-10% to total water consumption. This water footprint is especially critical in water-scarce areas where data centers may compete with municipal, agricultural, and industrial users (Li et al., 2023a; Han et al., 2024; Li et al., 2023b).

Water consumption is also affected by facility location and climate. Facilities in drier regions consume more water because they require more aggressive cooling, whereas facilities in temperate regions can consume less water through economizer-based cooling. Older legacy data centers report WUE values between 0.5 and 2.0 L/kWh (Li et al., 2023a; Jiang et al., 2025c; Liu & Mauter, 2021).

Environmental Impact

Water-resource depletion, carbon emissions, and local ecosystem impacts are among the environmental effects of traditional data centers. Carbon impacts depend on the carbon intensity of the electricity that supplies the facility (Guidi et al., 2024; Han et al., 2024; Li et al., 2023b).

Data centers located in regions with carbon-intensive electricity grids have higher greenhouse gas emissions. In the United States, data centers are estimated to contribute about 100-150 million metric tons of CO2 equivalent emissions annually, and traditional data centers account for a substantial share because of their higher PUE values. The CUE of traditional data centers ranges from 0.3 to 0.8 kg CO2e/kWh, depending on facility efficiency and grid carbon intensity (Qi et al., 2023b; Qi et al., 2023a; Hogade & Pasricha, 2024).

Water-resource impacts include water withdrawal, competition with other water users, and contamination risks associated with cooling towers. In water-scarce areas, data center water demand can affect municipal water supplies and surrounding ecosystems (Li et al., 2023a; Han et al., 2024; Li et al., 2023b).

Local environmental impacts should also consider land use, noise, and other facility-level effects because these factors contribute to the overall impact of data centers (Guidi et al., 2024; Han et al., 2024).

Economic Factors

The economic dimension of traditional data centers includes initial capital investment, recurring operating expenses, maintenance costs, and lifetime facility costs.

Conventional designs generally require lower initial investment because they rely on mature cooling, power-distribution, and facility-management technologies rather than AI-enabled control systems (Hogade & Pasricha, 2024; Li et al., 2025; Bahi & Ourici, 2025; Hasanbeig et al., 2022).

Water-consumption costs usually account for 5%-15% of total costs. Maintenance expenses are also relatively minor and manageable because established systems are already in place (Li et al., 2025; Bahi & Ourici, 2025).

The major economic disadvantage of older facilities is their comparatively higher costs of operation. Data centers that operate at PUE values of 1.6-1.8 may, for instance, find their energy costs 60-80% higher than facilities that operate in the range of 1.1-1.2 due to the enhanced consumption of energy by inefficient support systems. While incremental efficiencies typically produce a payback period in the range of 2-4 years, the improvements are implemented rather slowly due mainly to infrastructure inertia, lack of funding, and the prospect of operational challenges (Hogade & Pasricha, 2024; Li et al., 2025; Bahi & Ourici, 2025).

AI-DRIVEN DATA CENTERS: ARCHITECTURE & OPTIMIZATION

Architecture Summary

AI/ML-powered data centers embed artificial intelligence and machine learning technologies into facility infrastructure to reduce resource consumption. These designs retain the traditional IT, cooling, and power-distribution layers while adding an AI Optimization and Control Layer that monitors operating conditions and supports automated decision-making (Qi et al., 2023b; Qi et al., 2023a; Tuli et al., 2021).

The AI Optimization Layer functions as an intelligent control unit that continuously monitors, analyzes, and improves the performance of IT equipment, cooling systems, and power systems (Hogade & Pasricha, 2024; Naug et al., 2025). It operates through sensors, actuators, and control interfaces integrated with existing infrastructure (Tuli et al., 2021; Jiang et al., 2025a).

Intelligent workload management and dynamic resource allocation are features of the IT equipment as well. The consumption of idle power is reduced through the real-time prediction and control of server utilization, as well as the adjustment of power consumption based on the workload. Each of these is achieved through the application of machine learning to workload patterns (Tuli et al., 2021; Saxena & Singh, 2022). AI-driven optimizations in power distribution systems also add value to UPS operation streamlining, load balancing, and renewable power interconnections. Autonomous systems adjust to power demand forecasts, synergize with renewable power, and optimize battery control to smooth charging (Hogade & Pasricha, 2024; Li et al., 2025; Bahi & Ourici, 2025).

Figure 3 presents the AI-driven data center architecture, highlighting how the optimization layer interacts with conventional IT, cooling, and power systems to support predictive control and dynamic resource allocation.

Figure 3. AI-Driven Data Center Architecture with Optimization Layer

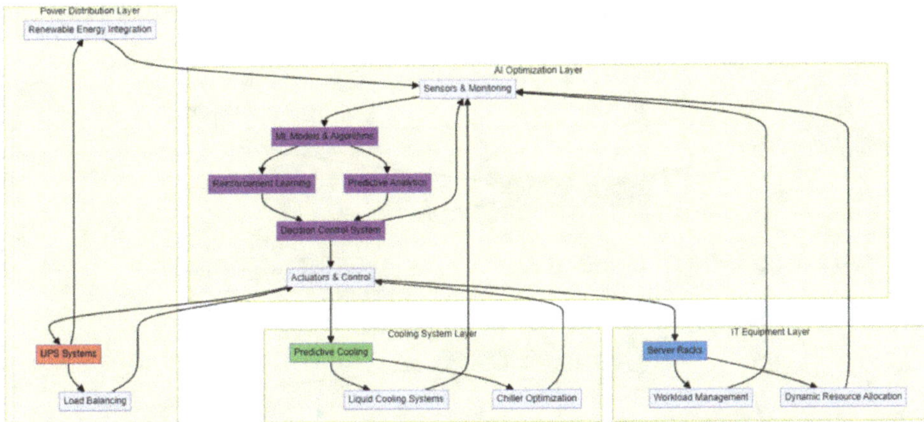

AI Optimization Technologies

AI-enabled data centers tackle optimization using predictive modeling, multi-objective optimization, reinforcement learning, and automated control techniques. These approaches use hybrid evolutionary learning and Pareto-based optimization to balance energy, cooling, carbon, and operational goals (Qi et al., 2023b; Qi et al., 2023a). HyperLoad is an example of a system that integrates large language models with cross-modal forecasting by using temperature, humidity, workload, and weather data (Jiang et al., 2025a). Liquid-cooling systems also use reinforcement learning; LC-Opt, for example, dynamically optimizes heat-exchanger control, flow rate, and pump operation (Naug et al., 2025).

AI-based predictive maintenance improves the scheduling of equipment maintenance, reduces unplanned downtime, and extends operational lifetimes (Tuli et al., 2021). Workload optimization includes dynamic resource allocation, virtual machine placement, and autoscaling. In geographically distributed data centers, AI can analyze workload patterns, resource availability, environmental conditions, and energy prices to support optimized placement decisions (Hogade & Pasricha, 2024; Li et al., 2023b). Figure 4 illustrates how workload, environmental, and system-state data are processed by the AI layer to guide resource-allocation decisions.

Figure 4. AI Optimization Workflow for Cooling and Resource Allocation

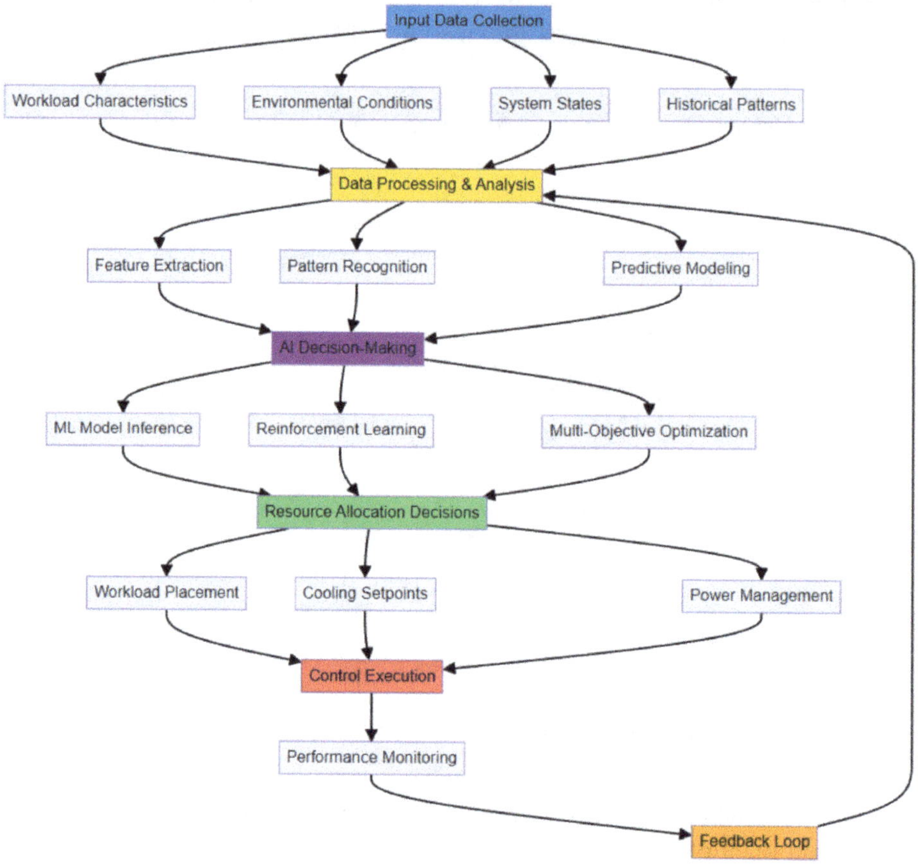

Improvements of Water and Energy Efficiency

AI-driven optimization has produced measurable improvements in energy and water-resource management. Compared with traditional facilities, AI-driven data centers can reduce energy use by approximately 20-35% and improve PUE from 1.4-1.8 to about 1.1-1.3 (Qi et al., 2023b; Qi et al., 2023a; Naug et al., 2025; Kannan et al., 2025). These gains are mainly associated with predictive cooling, intelligent workload management, and reduced power-distribution losses (Tuli et al., 2021; Kannan et al., 2025; Jiang et al., 2025a).

Predictive cooling and AI-based energy-management methods can reduce cooling energy by approximately 15-25% compared with conventional fixed-setpoint approaches (Naug et al., 2025; Jiang et al., 2025a). These systems cool only the zones and thermal loads that require intervention, and reinforcement learning can further improve liquid-cooling control (Sarkar et al., 2025; Naug et al., 2025).

AI also supports renewable-energy integration by forecasting renewable availability and scheduling workloads when cleaner energy is available (Li et al., 2025; Bahi & Ourici, 2025). Energy-management systems can balance renewable generation, battery storage, and workload demand to reduce grid dependence and lower carbon emissions.

In operation, AI-enabled data centers can achieve lower CUE values of approximately 0.1-0.3 kg CO2e/kWh, representing a substantial improvement over less optimized facilities (Qi et al., 2023b; Qi et al., 2023a; Hogade & Pasricha, 2024).

Challenges and Trade-offs

AI-optimized data centers have made progress toward improved efficiency. However, they still face water, energy, and heat-management challenges at the optimization and high-density computing levels (Li et al., 2023a; Jiang et al., 2025b; Singh et al., 2024; Jegham et al., 2025).

Large language models (LLMs) and AI workloads remain resource intensive. High-density AI accelerators can increase water, energy, and carbon impacts because they require additional cooling (Li et al., 2023a; Jiang et al., 2025b; Singh et al., 2024; Jegham et al., 2025). Therefore, AI facilities should be evaluated not by PUE alone but through PUE, WUE, CUE, and workload-related overhead.

AI data centers require greater upfront capital investment, often 20%-40% higher than traditional centers (Hogade & Pasricha, 2024; Li et al., 2025; Bahi & Ourici, 2025). These upfront costs are driven by AI computing infrastructure, sensors, and smart cooling and control systems. Larger hyperscale enterprises are more likely to recover these costs over the long term.

AI control systems increase operational complexity and require a combination of machine learning, facilities engineering, and model maintenance. Continuous recalibration and updating are necessary to keep decisions optimized and avoid model drift. Therefore, operational reliability and governance are as important as algorithmic performance for achieving sustainability outcomes (Tuli et al., 2021; Naug et al., 2025; Qi et al., 2023a).

The rebound effect, also known as Jevons' paradox, occurs when efficiency gains reduce operating costs and thereby increase overall demand for computation. This remains a concern for AI data centers (Luccioni et al., 2025).

Because of this, the value of optimization should be considered relative to computational cost. Self-optimizing systems can reduce cooling and power-distribution losses, but model training, inference, monitoring, and update costs may offset some gains if they are not managed carefully (Li et al., 2023a; Jiang et al., 2025b; Singh et al., 2024; Jegham et al., 2025).

COMPARATIVE ANALYSIS: ECONOMIC AND ENVIRONMENTAL ASPECTS

Energy Consumption Analysis

Under comparable assumptions, AI-enabled designs are generally more energy efficient. Traditional data centers usually report PUE values of 1.4-1.8, while AI-enabled facilities that use predictive cooling and intelligent control can report PUE values of 1.1-1.3. This indicates an estimated 20%-35% improvement in facility-level energy efficiency (Qi et al., 2023b; Qi et al., 2023a; Naug et al., 2025; Kannan et al., 2025).

In traditional facilities, IT equipment accounts for about 40%-50% of total facility energy consumption. Cooling accounts for 30%-40%, and power distribution accounts for 10%-20% (Kannan et al., 2025; Ajibola et al., 2021). AI-enabled facilities reduce energy consumption through predictive cooling, improved workload management, and better power management. These methods can save 15%-25% of cooling-related energy consumption (Naug et al., 2025; Jiang et al., 2025a).

For example, a facility with a 10 MW IT load and a PUE of 1.6 uses 16 MW of total facility energy. If the PUE is 1.2, it uses about 12 MW of total facility energy. This represents a 25% reduction in total facility energy (Qi et al., 2023b; Qi et al., 2023a; Kannan et al., 2025).

Carbon impacts can also decrease when workloads are managed to reduce carbon intensity and when cleaner electricity is used. Under these conditions, CUE can improve from 0.3-0.8 to 0.1-0.3 kg CO_2e/kWh.

Figure 5 shows this comparison using a 10 MW IT-load scenario. The figure shows that the main efficiency gain comes from reducing energy used for cooling and overhead systems rather than from reducing the IT workload itself.

Figure 5. Energy Consumption Comparison under a 10 MW IT-Load Scenario

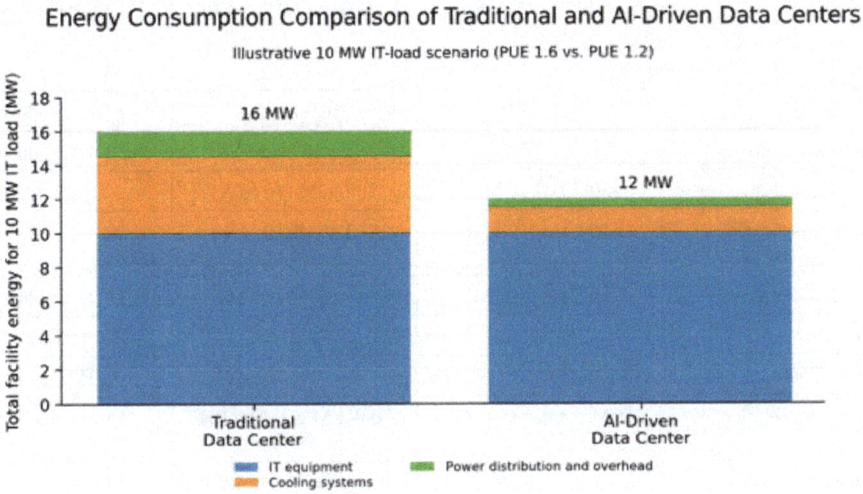

Economic Comparison

The economic comparison shows a trade-off between a greater initial financial hurdle and lower operating costs. Although AI-related facilities require more capital for smart control, advanced cooling, sensors, and AI resources, these costs may be recovered through lower energy use, reduced cooling demand, and improved operational efficiency (Hogade & Pasricha, 2024; Li et al., 2025; Bahi & Ourici, 2025).

For a representative 10 MW facility, a traditional data center may require $50-70 million, whereas an AI data center may require $60-90 million. This 20%-40% capital premium reflects additional costs for advanced cooling, control, monitoring, and AI-management systems (Hogade & Pasricha, 2024; Li et al., 2025; Bahi & Ourici, 2025).

AI data centers may achieve positive economic results over a 10-year span despite higher capital requirements. Traditional data centers may have a TCO in the range of $200-320 million, while AI data centers may be in the $160-270 million range, creating 15%-20% lifecycle-cost savings (Hogade & Pasricha, 2024; Li et al., 2025; Bahi & Ourici, 2025). Table 3 summarizes the environmental and economic metrics.

Table 3. Comprehensive Comparison of Traditional and AI-Driven Data Centers

Metric	Traditional Data Center	AI-Driven Data Center	Improvement
PUE	1.4 - 1.8	1.1 - 1.3	20-35% reduction
WUE (L/kWh)	0.5 - 2.0	0.3 - 1.5	15-25% reduction
CUE (kg CO2e/kWh)	0.3 - 0.8	0.1 - 0.3	30-40% reduction
Initial Capital Investment	$50-70M (10 MW)	$60-90M (10 MW)	20-40% higher
Annual Operational Expenses	$15-25M (10 MW)	$10-18M (10 MW)	20-40% reduction
10-Year TCO	$200-320M (10 MW)	$160-270M (10 MW)	15-20% reduction
ROI	Baseline	Positive	3-5 year payback
Energy Consumption	14-18 MW (10 MW IT)	11-13 MW (10 MW IT)	20-35% reduction
Water Consumption	1-5 M gal/day (10 MW)	0.8-4 M gal/day (10 MW)	15-25% reduction
Carbon Emissions	0.3-0.8 kg CO2e/kWh	0.1-0.3 kg CO2e/kWh	30-40% reduction

LIFECYCLE ASSESSMENT

A lifecycle assessment summarizes the phases of a facility from construction through operation to decommissioning. Although construction, manufacturing, and disposal phases of a facility contribute to the overall footprint, impacts of facility operation predominate. The facility consumes electricity, requires cooling, and uses water throughout its operational life.

During construction, embodied energy and water are associated with building materials, cooling infrastructure, electrical systems, servers, networking equipment, and installation. AI-driven facilities may require more advanced sensors and high-density infrastructure, but over a 10-15 year lifetime, operational impacts remain dominant.

During operation, AI-driven facilities can reduce energy, water, and carbon impacts through predictive cooling, workload optimization, and improved resource allocation. These gains are strongest when AI optimization is combined with renewable energy integration, carbon-aware workload placement, and water-efficient cooling systems.

Decommissioning impacts are broadly similar across both architectures, although material recovery, component reuse, and recycling can reduce end-of-life impacts when circular-economy practices are applied (Guidi et al., 2024; Han et al., 2024).

Across the full life cycle, AI-driven facilities may reduce total emissions by approximately 25-35% when operational savings outweigh embodied impacts. For a 10 MW facility operating over 15 years, this can correspond to an estimated reduction of 50,000-100,000 metric tons of CO_2 equivalent (Guidi et al., 2024; Hogade & Pasricha, 2024). Lifecycle water impacts show similar trends, with AI-driven facilities potentially reducing total water use by 20-30% when water-aware cooling and workload-management strategies are implemented (Li et al., 2023a; Jiang et al., 2025c; Guidi et al., 2024).

From an economic perspective, operational savings can outweigh higher initial cost. The NPV of AI-driven facilities can be 15-25% higher over a 15-year horizon when reductions in energy, water, maintenance, and carbon-related costs are included. When the deployment and operation of AI-driven systems are controlled, these systems can outperform traditional systems at every phase (Guidi et al., 2024; Hogade & Pasricha, 2024; Han et al., 2024; Li et al., 2025; Bahi & Ourici, 2025).

POLICY FRAMEWORK AND RECOMMENDATIONS

Policy Gaps and Challenges

Current policy frameworks tend to have important gaps in sustainability reporting. PUE reporting is mandatory in some jurisdictions, voluntary in others, and absent elsewhere. WUE reporting, water-consumption caps, and local water-stress assessments are even less developed (Jiang et al., 2025c), reducing comparison and accountability.

Carbon policy is also uneven because many frameworks do not require facility-level CUE reporting or disclosure of emissions associated with AI workloads, AI-intensive computation, cooling, and workload shifting (Qi et al., 2023b; Guidi et al., 2024; Hogade & Pasricha, 2024).

The uneven distribution of AI workloads introduces additional governance issues because many current frameworks do not clearly distinguish conventional digital infrastructure from AI facilities that use high-density computing, intensive cooling, and variable training and inference workloads (Li et al., 2023a; Jiang et al., 2025b; Ebert et al., 2024).

The gaps in policy are clearly reflected in a lack of frameworks for implementation, enforcement, and the structuring of compliance incentives. Most of the voluntary mechanisms for compliance lack sufficient accountability mechanisms (Ebert et al., 2024).

There is also significant disparity in economic incentives across the globe, resulting in uneven competition (Hogade & Pasricha, 2024; Li et al., 2025; Bahi & Ourici, 2025). The rebound effect (Jevons' paradox) is a complicating factor (Luccioni et al., 2025). The absence of harmonized lifecycle assessment standards leads to a fragmented assessment ecosystem (Guidi et al., 2024; Han et al., 2024).

Policy Recommendations

The situation warrants the provision of more specific policies. In the first instance, more prescriptive standardized reporting frameworks must be developed, whereby facilities with an excess of 1 MW IT capacity will be mandated to publicly report the metrics of PUE, WUE, and CUE on an annual basis (Ebert et al., 2024; Kannan et al., 2025). Reporting will be in accordance with the standards set in ISO/IEC 30134, and the report will be made available in a publicly accessible database.

In the second instance, there needs to be more stringent regulation around the reporting of water usage, such that regulations on WUE must be complemented with ceilings on water usage with respect to the local water supply. In addition, facilities located in areas experiencing water scarcity must be subject to more stringent requirements, such as the imposition of mandatory water recycling and the use of closed-loop systems. It is also necessary to provide greater incentives for the use of the WaterWise frameworks. Geographical load balancing can be used to determine and avoid areas of greater water stress for adaptive workloads (Jiang et al., 2025c; Li et al., 2023b).

Incentive structures must be designed economically to encourage the rapid integration of systems with renewable energy and AI-driven optimization. Investment in capital upfront can be financed through grants and tax incentives. In addition to the frameworks, incentive structures that capture the superior values of PUE, WUE, and CUE should be implemented.

Finally, policies must include the rebound effect and the product's full lifecycle. Policies should set limits on total resource consumption rather than focusing only on efficiency. A complete analysis should cover every stage of the lifecycle, and circular-economy principles should be included (Guidi et al., 2024; Han et al., 2024).

DISCUSSION

Key Findings Synthesis

The findings show that AI-enabled data centers have the potential to deliver significant environmental benefits through optimization of cooling, management of workloads, and control of the distribution of power. AI-enabled centers can have a PUE of 1.1-1.3 compared to 1.4-1.8 for traditional centers, a 15-25% reduction in the use of water, and a reduction in the carbon intensity of 0.3-0.8 kg CO2e/kWh to values of 0.1-0.3 kg CO2e/kWh (Qi et al., 2023b; Qi et al., 2023a; Li et al., 2023a; Jiang et al., 2025c; Hogade & Pasricha, 2024; Naug et al., 2025; Kannan et al., 2025). These findings are consistent with previous studies of multi-objective optimization and the use of predictive cooling. These findings also support the integrated consideration of PUE with WUE and CUE.

The economic model results also inform managerial decisions. Early adoption of AI infrastructure is estimated to increase capital investment by 20%-40%. Despite this, annual operational expenses are estimated to decrease by 20%-40%, with a 15%-20% reduction in 10-year total cost of ownership (TCO) and an expected payback period of 3-5 years (Hogade & Pasricha, 2024; Li et al., 2025; Bahi & Ourici, 2025). Given these predictions, investments in predictive cooling, reinforcement learning control, renewable integration, and intelligent resource allocation are justified. However, training, inference, monitoring, and rebound effects may limit total sustainability improvements if efficiency gains are offset by increased computational demand (Li et al., 2023a; Jiang et al., 2025b; Singh et al., 2024; Jegham et al., 2025; Luccioni et al., 2025).

Limitations

The analysis presented in this work has several limitations. First, it is based on public industry reports, benchmarked estimates, academic studies, and outline evidence rather than custom operational data. Second, the analysis does not account for extreme climate events or emergency operating conditions; instead, it focuses on steady-state system performance (Kannan et al., 2025; Jiang et al., 2025a). Third, the economic assumptions are standardized and do not account for differences across markets and regulatory environments (Hogade & Pasricha, 2024; Li et al., 2025; Bahi & Ourici, 2025). Fourth, the analysis is relative to facilities greater than 1 MW and may therefore be less relevant to smaller facilities or edge computing deployments. Finally, because the analysis is based on current technology, the findings may need revision as AI hardware, cooling systems, and sustainability-reporting standards evolve.

Directions for Future Research

AI-enabled data center performance demands research that prioritizes longitudinal, empirical studies that measure PUE, WUE, CUE, the cost of operations, and the cost and burden of workload over time (Qi et al., 2023b; Qi et al., 2023a; Hogade & Pasricha, 2024). More research is needed to determine the ways that training and performing inference with AI models harm the environment, and how to mitigate these effects (Li et al., 2023a; Jiang et al., 2025b; Singh et al., 2024; Jegham et al., 2025). Research is needed on rebound effects to determine how the increase in the efficiency of computational tasks can be transformed into the absolute reduction of the demand for resources (Luccioni et al., 2025).

Further research on AI optimization for smaller facilities and edge computing would broaden the applicability of this field (Tuli et al., 2021; Panteleaki et al., 2025; Schmitz et al., 2025). Studies combining renewable energy with AI-driven workload optimization could improve sustainability outcomes (Li et al., 2025; Bahi & Ourici, 2025). Lifecycle-impact studies beyond the operational phase would strengthen environmental assessment (Guidi et al., 2024; Han et al., 2024). Federated learning and distributed optimization may also help address scalability challenges (Liu et al., 2025), while policy-effectiveness studies can support evidence-based regulation (Ebert et al., 2024).

CONCLUSION

Summary of Findings

This comparative study shows that AI-driven data center architectures can improve environmental and economic performance when optimization is applied to cooling, workload management, power distribution, and resource allocation. AI-driven facilities can improve PUE from approximately 1.4-1.8 to 1.1-1.3, corresponding to a 20-35% improvement in facility-level energy efficiency. They can also reduce water consumption by approximately 15-25% (Li et al., 2023a; Jiang et al., 2025c) and lower carbon intensity by 30-40%, with CUE improving from 0.3-0.8 to 0.1-0.3 kg CO2e/kWh (Qi et al., 2023b; Qi et al., 2023a; Hogade & Pasricha, 2024).

The economic analysis shows that AI-driven facilities may require 20-40% higher initial capital investment, but they can also reduce annual operating costs by 20-40% (Li et al., 2025; Bahi & Ourici, 2025). Within a decade, AI-enhanced infrastructure may provide a total cost of ownership that is 15-20% lower, with a payback period of 3-5 years. Training costs, inference costs, and rebound effects remain important considerations for AI-driven facilities.

Contribution to Knowledge

This chapter consolidates the available literature on sustainable data centers and, to the best of the author's knowledge, is among the first to incorporate environmental, economic, and policy dimensions into a single comparative analysis. It identifies comparison ranges for PUE, WUE, and CUE (Qi et al., 2023b; Qi et al., 2023a; Li et al., 2023a). It also connects environmental data with TCO, ROI, and payback periods, arguing that sustainability can be economically reasonable when operational cost reductions exceed the initial investment (Hogade & Pasricha, 2024; Li et al., 2025; Bahi & Ourici, 2025). The chapter also contributes to policy-level sustainability assessment by addressing regulatory gaps in reporting, water use, carbon accountability, and AI workloads (Li et al., 2023a; Jiang et al., 2025c; Hogade & Pasricha, 2024; Ebert et al., 2024; Bahi & Ourici, 2025). The discussion of lifecycle assessment and rebound effects supports a more complete assessment of AI infrastructure (Guidi et al., 2024; Han et al., 2024; Luccioni et al., 2025).

Practical Implications

The results suggest important practical implications. AI-based optimization technologies are likely to yield a positive net present value in about 3-5 years for data center operators (Hogade & Pasricha, 2024; Li et al., 2025; Bahi & Ourici, 2025). Operators should adopt adaptive and predictive cooling, intelligent load balancing, renewable-energy integration, and geographic load balancing (Qi et al., 2023b; Qi et al., 2023a; Jiang et al., 2025c; Hogade & Pasricha, 2024; Li et al., 2023b; Naug et al., 2025).

Policymakers should set minimum reporting standards for PUE, WUE, and CUE, and water-consumption regulation should be strengthened in water-scarce regions (Li et al., 2023a; Jiang et al., 2025c; Ebert et al., 2024; Kannan et al., 2025). Policy should combine economic measures, carbon accountability, and sustainability-performance requirements to promote AI-based infrastructure deployment (Qi et al., 2023b; Hogade & Pasricha, 2024; Ebert et al., 2024; Li et al., 2025; Bahi & Ourici, 2025).

For technology companies, opportunities lie in AI-based optimization technologies that reduce cooling demand and resource consumption while making computing more reliable, scalable, and efficient .

Final Recommendations

Based on the analysis, data center operators should plan AI-led optimization implementations carefully. Strategies should focus on reducing computational overhead, geographic load balancing, renewable-energy integration, and water-aware workload management (Jegham et al., 2025). Policymakers should address current policy gaps through mandatory sustainability reporting, stricter water-use policies, carbon-emissions frameworks, and economic incentives for low-carbon infrastructure (Li et al., 2023a; Jiang et al., 2025c; Hogade & Pasricha, 2024; Ebert et al., 2024; Bahi & Ourici, 2025; Kannan et al., 2025).

Because optimization rebounds can offset efficiency gains, policy design should focus on total resource consumption rather than efficiency alone (Luccioni et al.,

2025; Ebert et al., 2024). Technology vendors should develop more efficient AI-led algorithms and hardware, together with federated learning and distributed optimization, to reduce computational and operational burdens (Singh et al., 2024; Jegham et al., 2025; Liu et al., 2025). Environmental stakeholders should also consider total resource impacts, including water, energy, carbon, and lifecycle effects .

More detailed lifecycle assessment is required to improve transparency in embodied energy, training-data infrastructure, model-training impacts, and associated water footprints. Further studies are also needed on longitudinal performance, computational burden, AI optimization for small-scale facilities, renewable integration, lifecycle impacts, and policy effectiveness.

This chapter shows that, when energy, water, carbon, economic, lifecycle, and policy dimensions are considered together, AI-based data center design has the potential to support more sustainable digital infrastructure (Li et al., 2023a; Jiang et al., 2025c; Qi et al., 2023b; Ebert et al., 2024).

REFERENCES

Ajibola, O. O., El-Gorashi, T. E. H., & Elmirghani, J. M. H. (2021). Energy-efficient placement of workloads in composable data center networks. arXiv. https://arxiv.org/abs/2101.08244

Alahmed, A. S., Botterud, A., Amin, S., & Al-Awami, A. T. (2025). Watts and drops: Co-scheduling power and water in desalination plants. arXiv. https://arxiv.org/abs/2509.19243

Amati, G., Turisini, M., & Acquaviva, A. (2024). Energy efficiency: A Lattice Boltzmann study. arXiv. https://arxiv.org/abs/2406.11498

Bahi, A., & Ourici, A. (2025). Deep reinforcement learning for real-time green energy integration in data centers. arXiv. https://arxiv.org/abs/2507.21153

Becchetti, L., Solferino, N., & Tessitore, M. E. (2025). The sustainable future is now: A dynamic model to advance investments in PV and energy storage. arXiv. https://arxiv.org/abs/2503.07131 DOI: 10.21203/rs.3.rs-6087030/v1

Di Martino, M., Linke, P., & Pistikopoulos, E. N. (2023). Towards optimal energy-water supply system operation for agricultural and metropolitan ecosystems. arXiv. https://arxiv.org/abs/2303.13599

Ebert, K., Alder, N., Herbrich, R., & Hacker, P. (2024). AI, climate, and regulation: From data centers to the AI Act. arXiv. https://arxiv.org/abs/2410.06681 DOI: 10.2139/ssrn.4980340

Goodarzi, M., & Li, Q. (2022). Security-constrained optimal operation of energy-water nexus based on a fast contingency filtering method. arXiv. https://arxiv.org/abs/2201.10785 DOI: 10.1109/PESGM48719.2022.9916824

Guidi, G., Dominici, F., Gilmour, J., Butler, K., Bell, E., Delaney, S., & Bargagli-Stoffi, F. J. (2024). Environmental burden of United States data centers in the artificial intelligence era. arXiv. https://arxiv.org/abs/2411.09786

Han, Y., Wu, Z., Li, P., Wierman, A., & Ren, S. (2024). The unpaid toll: Quantifying the public health impact of AI. arXiv. https://arxiv.org/abs/2412.06288

Hasanbeig, H., Kroening, D., & Abate, A. (2022). LCRL: Certified policy synthesis via logically-constrained reinforcement learning. arXiv. https://arxiv.org/abs/2209.10341 DOI: 10.1007/978-3-031-16336-4_11

He, Q. (2025). A unified metric architecture for AI infrastructure: A cross-layer taxonomy integrating performance, efficiency, and cost. arXiv. https://arxiv.org/abs/2511.21772 DOI: 10.2139/ssrn.5808163

Hogade, N., & Pasricha, S. (2024). Game-theoretic deep reinforcement learning to minimize carbon emissions and energy costs for AI inference workloads in geo-distributed data centers. arXiv. https://arxiv.org/abs/2404.01459

Jegham, N., Abdelatti, M., Elmoubarki, L., & Hendawi, A. (2025). How hungry is AI? Benchmarking energy, water, and carbon footprint of LLM inference. arXiv. https://arxiv.org/abs/2505.09598

Jiang, H., Qu, B., Zhu, J., Zeng, F., Lin, X., & Zhong, W. (2025a). HyperLoad: A cross-modality enhanced large language model-based framework for green data center cooling load prediction. arXiv. https://arxiv.org/abs/2512.19114

Jiang, Y., Kanakagiri, R., Roy, R. B., & Tiwari, D. (2025b). ThirstyFLOPS: Water footprint modeling and analysis toward sustainable HPC systems. arXiv. https://arxiv.org/abs/2510.00471 DOI: 10.1145/3712285.3759804

Jiang, Y., Roy, R. B., Kanakagiri, R., & Tiwari, D. (2025c). WaterWise: Co-optimizing carbon and water footprints toward environmentally sustainable cloud computing. arXiv. https://arxiv.org/abs/2501.17944 DOI: 10.1145/3710848.3710891

Kannan, D. D., Trivedi, A., & Srinivasan, D. (2025). A bidirectional gated recurrent unit model for PUE prediction in data centers. arXiv. https://arxiv.org/abs/2512.20161 DOI: 10.1109/IJCNN64981.2025.11227238

Li, P., Yang, J., Islam, M. A., & Ren, S. (2023a). Making AI less 'thirsty': Uncovering and addressing the secret water footprint of AI models. arXiv. https://arxiv.org/abs/2304.03271

Li, P., Yang, J., Wierman, A., & Ren, S. (2023b). Towards environmentally equitable AI via geographical load balancing. arXiv. https://arxiv.org/abs/2307.05494

Li, S., Tong, L., & Mount, T. D. (2025). Energy management for renewable-colocated artificial intelligence data centers. arXiv. https://arxiv.org/abs/2507.08011

Liu, J., Du, L., Li, Y., Liu, R.-P., Teng, F., & Hou, F. Y. (2025). Adaptive federated learning to optimize integrated flows in cyber-physical data centers. arXiv. https://arxiv.org/abs/2511.00623

Liu, Y., & Mauter, M. S. (2021). Marginal energy intensity of water supply. arXiv. https://arxiv.org/abs/2102.09753

Luccioni, A. S., Strubell, E., & Crawford, K. (2025). From efficiency gains to rebound effects: The problem of Jevons' paradox in AI's polarized environmental debate. arXiv. https://arxiv.org/abs/2501.16548 DOI: 10.1145/3715275.3732007

Mabrouk, M. T., Majidniya, M., & Lacarrière, B. (2025). Optimal scheduling of energy and mass flows based on networked multi-carrier hubs formulation: a general framework. arXiv. https://arxiv.org/abs/2504.17341

Mitra, B., Datta, S., Kincic, S., Samaan, N., & Somani, A. (2023). Gaps in representations of hydropower generation in steady-state and dynamic models. arXiv. https://arxiv.org/abs/2311.03277

Naug, A., Sarkar, S., Tuli, S., & Buyya, R. (2025). LC-Opt: Benchmarking reinforcement learning and agentic AI for end-to-end liquid cooling optimization in data centers. arXiv. https://arxiv.org/abs/2511.00116

Obringer, R., Nateghi, R., Knee, J., Madani, K., & Kumar, R. (2023). Contemporary climate analogs project north-south polarization of urban water-energy nexus across US cities under warming climate. arXiv. https://arxiv.org/abs/2306.17050

Panteleaki, A. M., Paramanayakam, V., Pentsos, V., Karatzas, A., Tragoudas, S., & Anagnostopoulos, I. (2025). A vertical approach to designing and managing sustainable heterogeneous edge data centers. arXiv. https://arxiv.org/abs/2506.01712 DOI: 10.1109/ISVLSI65124.2025.11130313

Qi, S., Milojicic, D., Bash, C., & Pasricha, S. (2023a). MOSAIC: A multi-objective optimization framework for sustainable data center management. arXiv. https://arxiv.org/abs/2311.08583

Qi, S., Milojicic, D., Bash, C., & Pasricha, S. (2023b). SHIELD: Sustainable hybrid evolutionary learning framework for carbon, wastewater, and energy-aware data center management. arXiv. https://arxiv.org/abs/2308.13086 DOI: 10.1145/3634769.3634810

Quach, R., Wang, Y., Jahanshahi, A., Wong, D., & Kim, H. (2025). ECLIP: Energy-efficient and practical co-location of ML inference on spatially partitioned GPUs. arXiv. https://arxiv.org/abs/2506.12598 DOI: 10.1109/ISLPED65674.2025.11261793

Saheb, T., & Dehghani, M. (2021). Artificial intelligence for sustainable energy: A contextual topic modeling and content analysis. arXiv. https://arxiv.org/abs/2110.00828

Sarkar, S., Tuli, S., & Buyya, R. (2025). Hierarchical multi-agent framework for carbon-efficient liquid-cooled data center clusters. arXiv. https://arxiv.org/abs/2502.08337

Saxena, D., & Singh, A. K. (2022). A proactive autoscaling and energy-efficient VM allocation framework using online multi-resource neural network for cloud data center. arXiv. https://arxiv.org/abs/2212.01896

Schmitz, R., Flachsbarth, F., Plaga, L. S., Braun, M., & Härtel, P. (2025). Energy security and resilience: Reviewing concepts and advancing planning perspectives for transforming integrated energy systems. arXiv. https://arxiv.org/abs/2504.18396

Silva-Rodriguez, J., & Li, X. (2023). Decentralized micro water-energy co-optimization for small communities. arXiv. https://arxiv.org/abs/2310.01681

Singh, A., Patel, N. P., Ehtesham, A., Kumar, S., & Khoei, T. T. (2024). A survey of sustainability in large language models: Applications, economics, and challenges. arXiv. https://arxiv.org/abs/2412.04782

Tuli, S., Ilager, S., Ramamohanarao, K., & Buyya, R. (2021). HUNTER: AI-based holistic resource management for sustainable cloud computing. arXiv. https://arxiv.org/abs/2110.05529

Zhan, X., Zhu, X., Cheng, P., Hu, X., He, Z., Geng, H., Leng, J., Zheng, H., Liu, C., Hong, T., Liang, Y., Liu, Y., & Zhao, F. (2025). Data center cooling system optimization using offline reinforcement learning. arXiv. https://arxiv.org/abs/2501.15085

KEY TERMS AND DEFINITIONS

AI-driven data centers: Data centers that use artificial intelligence to optimize cooling, workload distribution, energy consumption, and predictive maintenance in real time.

Artificial Intelligence (AI) Optimization: The application of machine learning, predictive analytics, and reinforcement learning techniques to dynamically manage cooling, workload allocation, and resource utilization in data centers for improved efficiency.

Artificial intelligence (AI): Systems that continuously learn by imitating the human mind to perform tasks such as decision-making, pattern recognition, and process optimization.

Carbon Usage Effectiveness (CUE): A metric that measures carbon emissions per unit of IT equipment energy, expressed in kg CO_2e/kWh, used to evaluate the environmental impact of data center operations.

Data Center Sustainability: The balanced consideration of energy consumption, water usage, carbon emissions, economic viability, and regulatory compliance in the design and operation of data centers.

Energy efficiency: The ratio of useful output (e.g., computing work) to energy input, often improved through better cooling, power management, and hardware optimization.

Environmental impact: The effect of data center operations on the natural environment, including carbon emissions, water consumption, land use, and electronic waste.

Geographic Load Balancing: The strategic distribution of computational workloads across multiple data centers based on energy prices, carbon intensity, water availability, and renewable energy sources.

Lifecycle Assessment (LCA): A comprehensive evaluation of environmental impacts across all stages of a data center's life, from construction and operation to decommissioning and disposal.

Policy Framework: The collection of regulations, reporting standards, and incentives governing data center sustainability, including requirements for PUE, WUE, and CUE disclosure.

Power Usage Effectiveness (PUE): The ratio of total facility energy consumption to IT equipment energy consumption. A lower PUE indicates higher energy efficiency of the supporting infrastructure.

Sustainability policy: A formal set of guidelines and commitments adopted by an organization to reduce its environmental footprint, improve resource efficiency, and comply with regulatory standards.

Total Cost of Ownership (TCO): The complete cost of owning and operating a data center over its lifecycle, including initial capital investment and ongoing operational expenses.

Water Usage Effectiveness (WUE): A metric that quantifies water consumption per unit of IT equipment energy, expressed in liters per kWh, critical for assessing water footprint in data centers.

Compilation of References

Ajibola, O. O., El-Gorashi, T. E. H., & Elmirghani, J. M. H. (2021). Energy-efficient placement of workloads in composable data center networks. arXiv. https://arxiv.org/abs/2101.08244

Al Kez, D., Foley, A. M., Laverty, D., Del Rio, D. F., & Sovacool, B. (2022). Exploring the sustainability challenges facing digitalization and internet data centers. *Journal of Cleaner Production*, *371*, 133633. Advance online publication. DOI: 10.1016/j.jclepro.2022.133633

Alahmed, A. S., Botterud, A., Amin, S., & Al-Awami, A. T. (2025). Watts and drops: Co-scheduling power and water in desalination plants. arXiv. https://arxiv.org/abs/2509.19243

Alsouda, F., Bennett, N. S., Saha, S. C., Salehi, F., & Islam, M. S. (2023). Vapor compression cycle: A state-of-the-art review on cycle improvements, water and other natural refrigerants. *Cleanroom Technology*, *5*(2), 584–608. DOI: 10.3390/cleantechnol5020030

Amati, G., Turisini, M., & Acquaviva, A. (2024). Energy efficiency: A Lattice Boltzmann study. arXiv. https://arxiv.org/abs/2406.11498

Arizona v. California, 373 U.S. 546 (1963).

Bahi, A., & Ourici, A. (2025). Deep reinforcement learning for real-time green energy integration in data centers. arXiv. https://arxiv.org/abs/2507.21153

Bashroush, R., Rteil, N., Kenny, R., & Wynne, A. (2020). Optimizing server refresh cycles: The case for circular economy with an aging moore's law. *IEEE Transactions on Sustainable Computing*, *7*(1), 189–200. DOI: 10.1109/TSUSC.2020.3035234

Becchetti, L., Solferino, N., & Tessitore, M. E. (2025). The sustainable future is now: A dynamic model to advance investments in PV and energy storage. arXiv. https://arxiv.org/abs/2503.07131 DOI: 10.21203/rs.3.rs-6087030/v1

Ben-Amar, W., & Chelli, M. (2018). What drives voluntary corporate water disclosures? The effect of country-level institutions. *Business Strategy and the Environment, 27*(8), 1609–1622. DOI: 10.1002/bse.2227

Blackhurst, M., Hendrickson, C., & Vidal, J. S. I. (2010). Direct and indirect water withdrawals for U.S. industrial sectors. *Environmental Science & Technology, 44*(6), 2126–2131. DOI: 10.1021/es903147k PMID: 20141104

Bolón-Canedo, V., Morán-Fernández, L., Cancela, B., & Alonso-Betanzos, A. (2024). A review of green artificial intelligence: Towards a more sustainable future. *Neurocomputing, 565*, 128096. DOI: 10.1016/j.neucom.2024.128096

Borgesius, F. Z., Trilling, D., Moeller, J., Bodó, B., de Vreese, C. H., & Helberger, N. (2022). Should we worry about filter bubbles? *Internet Policy Review, 11*(1), 1–16. DOI: 10.14763/2022.1.1561

Bozkurt, A. (2025). AI's Thirst, AI's Heat, AI's Waste: Exposing the Hidden Environmental Impact of Every Artificial Intelligence Interaction. *Open Praxis, 17*(4), 638–647. DOI: 10.55982/openpraxis.17.4.1058

Brodie, P. (2023). Data infrastructure studies on an unequal planet. *Big Data & Society, 10*(1), 20539517231182402. Advance online publication. DOI: 10.1177/20539517231182402

Brundage, M., Avin, S., Clark, J., Toner, H., Eckersley, P., Garfinkel, B., Dafoe, A., Scharre, P., Zeitzoff, T., & Filar, B. (2018). The malicious use of artificial intelligence: Forecasting, prevention, and mitigation. *arXiv preprint arXiv:1802.07228.*

Bund für Umwelt und Naturschutz Deutschland v. Germany (Weser River Case), Case C-461/13 (2015).

Burritt, R. L., Christ, K. L., & Omori, A. (2016). Drivers of corporate water-related disclosure: Evidence from Japan. *Journal of Cleaner Production, 129*, 65–74. DOI: 10.1016/j.jclepro.2016.04.119

Cannon, C. E. B., Chu, E. K., Natekal, A., & Waaland, G. (2024). Institutional designs for procedural justice and inclusion in urban climate change adaptation. *Journal of Planning Education and Research, 46*(1), 1–18. DOI: 10.1177/0739456X241274579

Cao, K., Li, Z., Luo, H., Jiang, Y., Liu, H., Xu, L., Gao, P., & Liu, H. (2024). Comprehensive review and future prospects of multi-level fan control strategies in data centers for joint optimization of thermal management systems. *Journal of Building Engineering, 94*, 110021. Advance online publication. DOI: 10.1016/j.jobe.2024.110021

Cerquitelli, T., Nikolakis, N., Bethaz, P., Panicucci, S., Ventura, E., Macii, E., Andolina, A., Marguglio, K., Alexopoulos, K., Petrali, P., Pagani, A., van Wilgen, M., & Ippolito, M. (2020). Enabling predictive analytics for smart manufacturing through an IIoT platform. *IFAC-PapersOnLine*, *53*(3), 179–184. Advance online publication. DOI: 10.1016/j.ifacol.2020.11.029

Chen, C., Ye, C., Ding, Y., Xiong, K., & Tang, X. (2026). Exploring the Demand Response Potential of Liquid-Cooled Data Centers. *Applied Energy*, *412*, 127679. Advance online publication. DOI: 10.1016/j.apenergy.2026.127679

Chouksey, A., Rajan, A. K., Gurjar, V., Tiwari, R., & Mishra, P. K. (2026). The green paradox: The climate, environmental, and sustainability implications of artificial intelligence. *Global Environmental Change Advances*, *6*, 100029. DOI: 10.1016/j.gecadv.2025.100029

Clarke, T. (2025). Sustainability in international public policy and governance.

Corigliano, O., Algieri, A., & Fragiacomo, P. (2024). Turning Data Center Waste Heat into Energy: A Guide to Organic Rankine Cycle System Design and Performance Evaluation. *Applied Sciences (Basel, Switzerland)*, *14*(14), 6046. Advance online publication. DOI: 10.3390/app14146046

Crawford, K. (2021). *Atlas of AI: Power, politics, and the planetary costs of artificial intelligence*. Yale University Press.

d'Orgeval, A., Sheehan, S., Avenas, Q., Assoumou, E., & Sessa, V. (2026). Generative AI impact assessment through a life cycle analysis of multiple data center typologies. *Applied Energy*, *406*, 127288. Advance online publication. DOI: 10.1016/j.apenergy.2025.127288

da Silva, L. A., Vasconcelos, E. S., de Paiva, L. F. R., Collaço, M. H. d. V. R., Joaquim, W. M., de Lima, A. D., Goulart, C. S., & Teixeira, E. P. (2025). ENERGY CONSUMPTION OF DATA CENTERS AND THE PRIMARY ENERGY METRIC: GLOBAL ANALYSIS 2000–2024 AND PROJECTIONS UNTIL 2030. *Editora Impacto Científico*, 21-41.

Dayarathna, M., Wen, Y., & Fan, R. (2016). Data center energy consumption modeling: A survey. *IEEE Communications Surveys and Tutorials*, *18*(1), 732–794. DOI: 10.1109/COMST.2015.2481183

de Vries-Gao, A. (2025). The carbon and water footprints of data centers and what this could mean for artificial intelligence. *Patterns (New York, N.Y.)*, *7*(1), 101430. DOI: 10.1016/j.patter.2025.101430 PMID: 41583976

Dellapenna, J. W., & Gupta, J. (2020). *The evolution of the law and politics of water*. Springer.

Di Martino, M., Linke, P., & Pistikopoulos, E. N. (2023). Towards optimal energy-water supply system operation for agricultural and metropolitan ecosystems. arXiv. https://arxiv.org/abs/2303.13599

du Plessis, A. (2023). Water resources from a global perspective. In *South Africa's Water predicament: Freshwater's unceasing decline* (pp. 1–25). Springer. DOI: 10.1007/978-3-031-24019-5_1

Ebert, K., Alder, N., Herbrich, R., & Hacker, P. (2024). AI, climate, and regulation: From data centers to the AI Act. arXiv. https://arxiv.org/abs/2410.06681 DOI: 10.2139/ssrn.4980340

El Hanjri, M., Kabbaj, H., Kobbane, A., & Abouaomar, A. (2023). Federated learning for water consumption forecasting in smart cities. arXiv. /arXiv.2301.13036 DOI: 10.1109/ICC45041.2023.10279576

Farfan, J., & Lohrmann, A. (2023). Gone with the clouds: Estimating the electricity and water footprint of digital data services in Europe. *Energy Conversion and Management*, *290*, 117225. DOI: 10.1016/j.enconman.2023.117225

Farooq, M. B., Naveed, K., Khalid, F., Narayan, A. K., & Khudir, I. M. (2025). Examining the extent and quality of corporate water management disclosures in extremely high-water stress countries. Sustainability Accounting. *Management and Policy Journal*, *16*(3), 705–733. DOI: 10.1108/SAMPJ-01-2024-0054

Floridi, L., Cowls, J., Beltrametti, M., Chatila, R., Chazerand, P., Dignum, V., Luetge, C., Madelin, R., Pagallo, U., Rossi, F., Schafer, B., Valcke, P., & Vayena, E. (2018). AI4People—An ethical framework for a good AI society: Opportunities, risks, principles, and recommendations. *Minds and Machines*, *28*(4), 689–707. DOI: 10.1007/s11023-018-9482-5 PMID: 30930541

Food Marketing Institute v. Argus Leader Media, 588 U.S. ___ (2019).

Friends of the Earth, Inc. v. Laidlaw Environmental Services, 528 U.S. 167 (2000).

Fu, Z.-L., Cao, C., & Gao, F. (2025). Impact of artificial intelligence energy management technologies on commercial multi-energy consumption. *Energy*, *340*, 139247. Advance online publication. DOI: 10.1016/j.energy.2025.139247

Gao, J., Shao, C., Chen, S., & Zhang, X. (2021). Spatiotemporal evolution of sustainable development of China's provinces: A modelling approach. *Ecosystem Health and Sustainability*, *7*(1), 1965034. DOI: 10.1080/20964129.2021.1965034

Gao, P. X., Curtis, A. R., Wong, B., & Keshav, S. (2012). It's not easy being green. In *Proceedings of the ACM SIGCOMM Conference*. DOI: 10.1145/2342356.2342398

Gehring, J. C., & Cand, B. (2024). *The Sustainable Energy Imperative: A Future Generations Perspective on Technologies Leading the Clean Energy Transition*. The Global Youth Council on Science, Law & Sustainability.

Geissdoerfer, M., Savaget, P., Bocken, N. M. P., & Hultink, E. J. (2017). The circular economy: A new sustainability paradigm? *Journal of Cleaner Production, 143*, 757–768. DOI: 10.1016/j.jclepro.2016.12.048

Getches, D. H. (2017). *Water law in a nutshell* (5th ed.). West Academic Publishing.

Ghorbani Bam, P., Ochoa-Rodriguez, S., & Butler, D. (2025). Digital twin applications in the water sector: A review. *Water (Basel), 17*(20), 2957. DOI: 10.3390/w17202957

Goodarzi, M., & Li, Q. (2022). Security-constrained optimal operation of energy-water nexus based on a fast contingency filtering method. arXiv. https://arxiv.org/abs/2201.10785 DOI: 10.1109/PESGM48719.2022.9916824

Google. (2021). Water stewardship in data centers. https://sustainability.google/reports/google-water-stewardship

Guidi, G., Dominici, F., Gilmour, J., Butler, K., Bell, E., Delaney, S., & Bargagli-Stoffi, F. J. (2024). Environmental burden of United States data centers in the artificial intelligence era. arXiv. https://arxiv.org/abs/2411.09786

Haddout, S., & Kurniawan, T. A. (Eds.). (2026). *Handbook of Artificial Intelligence in Water Management: Towards Net Zero Emissions* (1st ed.). CRC Press., DOI: 10.1201/9781003547099

Haghdadi, N., Cizek, P., Hodgson, P., Tari, V., Rohrer, G., & Beladi, H. (2018). Effect of ferrite-to-austenite phase transformation path on the interface crystallographic character distributions in a duplex stainless steel. *Acta Materialia, 145*, 196–209. DOI: 10.1016/j.actamat.2017.11.057

Han, Y., Wu, Z., Li, P., Wierman, A., & Ren, S. (2024). The unpaid toll: Quantifying the public health impact of AI. arXiv. https://arxiv.org/abs/2412.06288

Hasanbeig, H., Kroening, D., & Abate, A. (2022). LCRL: Certified policy synthesis via logically-constrained reinforcement learning. arXiv. https://arxiv.org/abs/2209.10341 DOI: 10.1007/978-3-031-16336-4_11

He, Q. (2025). A unified metric architecture for AI infrastructure: A cross-layer taxonomy integrating performance, efficiency, and cost. arXiv. https://arxiv.org/abs/2511.21772 DOI: 10.2139/ssrn.5808163

Herrera, M., Xie, X., Menapace, A., Zanfei, A., & Brentan, B. M. (2025). Sustainable AI infrastructure: A scenario-based forecast of water footprint under uncertainty. *Journal of Cleaner Production, 526*, 146528. DOI: 10.1016/j.jclepro.2025.146528

Hintemann, R., & Hinterholzer, S. (2020). Energy consumption of data centers worldwide: Current development and future challenges. *Energy Efficiency*, *13*(7), 1365–1381. DOI: 10.1007/s12053-020-09874-9

Hiremath, R. B. (2024). AI-embedded data centres: Promoting sustainability and reducing water footprint. In *2024 First International Conference on Data, Computation and Communication (ICDCC)* (pp. 40–44). IEEE. DOI: 10.1109/ICDCC62744.2024.10961316

Hlabisa, S. (2025). The ecology of artificial intelligence: Energy, water, materials, and land limits of digital systems. *Carbon Neutral Systems.*, *1*(1), 19. DOI: 10.1007/s44438-025-00018-8

Hoekstra, A. Y. (2011a). *The water footprint assessment manual: Setting the global standard*. Routledge.

Hoekstra, A. Y., Chapagain, A. K., Aldaya, M. M., & Mekonnen, M. M. (2011b). *The Water Footprint Assessment Manual: Setting the Global Standard*. Earthscan.

Hoff, H. (2011). *Understanding the nexus: Background paper for the Bonn 2011 conference on the water, energy and food security nexus*. Stockholm Environment Institute.

Hogade, N., & Pasricha, S. (2024). Game-theoretic deep reinforcement learning to minimize carbon emissions and energy costs for AI inference workloads in geo-distributed data centers. arXiv. https://arxiv.org/abs/2404.01459

Hoover, F.-A., Meerow, S., Grabowski, Z. J., & McPhearson, T. (2021). Environmental justice implications of siting criteria in urban green infrastructure planning. *Journal of Environmental Policy and Planning*, *23*(5), 665–682. DOI: 10.1080/1523908X.2021.1945916

Illinois Central Railroad Co. v. Illinois, 146 U.S. 387 (1892).

International Energy Agency. (2022). *Data centers and data transmission networks*. IEA. https://www.iea.org/

International Energy Agency. (2024). *Electricity 2024: Analysis and forecast to 2026*. IEA.

International Energy Agency. (2025). *Energy and AI*. Author.

International Energy Agency. (2025). Energy and AI. https://www.iea.org/reports/energy-and-ai

International Energy Agency. (2026, April 16). Data centre electricity use surged in 2025, even with tightening bottlenecks driving a scramble for solutions. https://www.iea.org/news/data-centre-electricity-use-surged-in-2025-even-with-tightening-bottlenecks-driving-a-scramble-for-solutions

Islam, S., Smith, K., Klimes, M., & Salzberg, A. (Eds.). (2025). *Routledge Handbook of Water Diplomacy* (1st ed.). Routledge., DOI: 10.4324/9781003178439

Jegham, N., Abdelatti, M., Elmoubarki, L., & Hendawi, A. (2025). How hungry is AI? Benchmarking energy, water, and carbon footprint of LLM inference. arXiv. https://arxiv.org/abs/2505.09598

Jiang, F., Duan, C., & Chen, B. (2025). Facility-level energy-driven water footprint and scarcity implications of Chinese data centers: A bottom-up analysis and scenario-based projection. *Applied Energy*, *399*, 126522. DOI: 10.1016/j.apenergy.2025.126522

Jiang, H., Li, N., Zhao, C., & Shen, L. (2026). Research on return water temperature prediction model for casting cooling system based on deep learning. *Scientific Reports*, *16*(1), 4432. Advance online publication. DOI: 10.1038/s41598-025-34510-5 PMID: 41501340

Jiang, H., Qu, B., Zhu, J., Zeng, F., Lin, X., & Zhong, W. (2025a). HyperLoad: A cross-modality enhanced large language model-based framework for green data center cooling load prediction. arXiv. https://arxiv.org/abs/2512.19114

Jiang, Y., Kanakagiri, R., Roy, R. B., & Tiwari, D. (2025b). ThirstyFLOPS: Water footprint modeling and analysis toward sustainable HPC systems. arXiv. https://arxiv.org/abs/2510.00471 DOI: 10.1145/3712285.3759804

Jiang, Y., Roy, R. B., Kanakagiri, R., & Tiwari, D. (2025c). WaterWise: Co-optimizing carbon and water footprints toward environmentally sustainable cloud computing. arXiv. https://arxiv.org/abs/2501.17944 DOI: 10.1145/3710848.3710891

Jin, Y., Behrens, P., Tukker, A., & Scherer, L. (2019). Water use of electricity technologies: A global meta-analysis. *Renewable & Sustainable Energy Reviews*, *115*, 109391. DOI: 10.1016/j.rser.2019.109391

Jobin, A., Ienca, M., & Vayena, E. (2019). The global landscape of AI ethics guidelines. *Nature Machine Intelligence*, *1*(9), 389–399. DOI: 10.1038/s42256-019-0088-2

Jones, N. (2018). How to stop data centres from gobbling up the world's electricity. *Nature*, *561*(7722), 163–166. DOI: 10.1038/d41586-018-06610-y PMID: 30209383

Kannan, D. D., Trivedi, A., & Srinivasan, D. (2025). A bidirectional gated recurrent unit model for PUE prediction in data centers. arXiv. https://arxiv.org/abs/2512.20161 DOI: 10.1109/IJCNN64981.2025.11227238

Kollar, A., & Grady, C. (2025). The relationship between data centers and the climate is a systems challenge: A spatial analysis of United States data centers. *Environmental Research Communications*, *7*(11), 111005. DOI: 10.1088/2515-7620/ae193a

Koomey, J. (2011). *Growth in data center electricity use 2005 to 2010.* Analytics Press.

Kurniawan, T. A., & Anouzla, A. (Eds.). (2026). *Handbook of Digitalization and Big Data in the Water Sector: Emerging Opportunities and Challenges in Net Zero Emissions* (1st ed., Vol. 2). CRC Press., DOI: 10.1201/9781003536079

Kuzma, S., Bierkens, M. F. P., Lakshman, S., Luo, T., Saccoccia, L., Sutanudjaja, E. H., & Van Beek, R. (2023). *Aqueduct 4.0: Updated decision-relevant global water risk indicators.* World Resources Institute.

Lambert, K., & Luccioni, S. (2026). *From Cradle to Cloud: A Life Cycle Review of AI's Environmental Footprint.* ArXiv., DOI: 10.1145/3805689.3812401

Lambert, S., Caron, E., Lefevre, L., & Grivel, R. (2026). Consolidation of virtual machines to reduce energy consumption of data centers by using ballooning, sharing and swapping mechanisms. *Future Generation Computer Systems*, *174*, 107968. Advance online publication. DOI: 10.1016/j.future.2025.107968

Lannelongue, L., Grealey, J., & Inouye, M. (2021). Green algorithms: Quantifying the carbon footprint of computation. *Advanced Science (Weinheim, Baden-Württemberg, Germany)*, *8*(12), 2100707. DOI: 10.1002/advs.202100707 PMID: 34194954

Lee, U., Han, J., Elgowainy, A., & Wang, M. (2018). Regional water consumption for hydro and thermal electricity generation in the United States. *Applied Energy*, *210*, 661–672. DOI: 10.1016/j.apenergy.2017.05.025

Lei, N., Lu, J., Shehabi, A., & Masanet, E. (2025). The water use of data center workloads: A review and assessment of key determinants. *Resources, Conservation and Recycling*, *219*, 108310. DOI: 10.1016/j.resconrec.2025.108310

Leon, L., Chaplot, B., & Solomon, A. (2020). Water consumption forecasting using soft computing: A case study. *Water Science and Technology: Water Supply*, *20*(8), 3576–3584. DOI: 10.2166/ws.2020.273

Li, C., Shu, H., & Gu, X. (2024). Photoperiod Management in Farm Animal Husbandry: A Review. *Animals (Basel)*, *15*(4), 591–604. DOI: 10.3390/ani15040591 PMID: 40003072

Lin, J., Lin, W., Huang, H., Lin, W., & Li, K. (2024). Thermal Modeling and Thermal-Aware Energy Saving Methods for Cloud Data Centers: A Review. *IEEE Transactions on Sustainable Computing*, *9*(3), 571–590. DOI: 10.1109/TSUSC.2023.3346332

Li, P., Yang, J., Islam, M. A., & Ren, S. (2023a). Making AI less 'thirsty': Uncovering and addressing the secret water footprint of AI models. arXiv. https://arxiv.org/abs/2304.03271

Li, P., Yang, J., Islam, M. A., & Ren, S. (2025). Making ai less' thirsty'. *Communications of the ACM*, *68*(7), 54–61. DOI: 10.1145/3724499

Li, P., Yang, J., Wierman, A., & Ren, S. (2023b). Towards environmentally equitable AI via geographical load balancing. arXiv. https://arxiv.org/abs/2307.05494

Li, S., Tong, L., & Mount, T. D. (2025). Energy management for renewable-colocated artificial intelligence data centers. arXiv. https://arxiv.org/abs/2507.08011

Liu, J., Du, L., Li, Y., Liu, R.-P., Teng, F., & Hou, F. Y. (2025). Adaptive federated learning to optimize integrated flows in cyber-physical data centers. arXiv. https://arxiv.org/abs/2511.00623

Liu, Y., & Mauter, M. S. (2021). Marginal energy intensity of water supply. arXiv. https://arxiv.org/abs/2102.09753

Liu, Y., Wei, X., Xiao, J., Liu, Z., Xu, Y., & Tian, Y. (2020). Energy consumption and emission mitigation prediction based on data center traffic and PUE for global data centers. *Global Energy Interconnection*, *3*(3), 272–282. DOI: 10.1016/j.gloei.2020.07.008

Long, N. V., Vinh, T. N., Quyen, H. V., Duyen, N. T. M., Hoa, P. V., & Tuan, M. A. (2026). The Evolution of AI Chips: From Cloud to Edge, Powering the Future of Artificial Intelligence. In *Lecture Notes of the Institute for Computer Sciences, Social-Informatics and Telecommunications Engineering, LNICST* (Vol. 673 LNICST, pp. 188–217). DOI: 10.1007/978-3-032-14055-5_14

Luccioni, A. S., Strubell, E., & Crawford, K. (2025). From efficiency gains to rebound effects: The problem of Jevons' paradox in AI's polarized environmental debate. arXiv. https://arxiv.org/abs/2501.16548 DOI: 10.1145/3715275.3732007

Lundberg, S. M., & Lee, S.-I. (2017). A unified approach to interpreting model predictions. arXiv. https://arxiv.org/abs/1705.07874

M.C. Mehta v. Kamal Nath, (1997) 1 SCC 388.

Mabrouk, M. T., Majidniya, M., & Lacarrière, B. (2025). Optimal scheduling of energy and mass flows based on networked multi-carrier hubs formulation: a general framework. arXiv. https://arxiv.org/abs/2504.17341

Macknick, J., Newmark, R., Heath, G., & Hallett, K. C. (2012). Operational water consumption and withdrawal factors for electricity generating technologies: A review of existing literature. *Environmental Research Letters*, *7*(4), 045802. DOI: 10.1088/1748-9326/7/4/045802

Magotra, B., Malhotra, D., & Dogra, A. K. (2023). Correction to: Adaptive Computational Solutions to Energy Efficiency in Cloud Computing Environment Using VM Consolidation. *Archives of Computational Methods in Engineering*, *30*(5), 3485. DOI: 10.1007/s11831-022-09868-8

Manoharan, G., Ashtikar, S. P., & Nivedha, M. (2024). Unveiling the critical role of artificial intelligence in energy management. In *AI applications for clean energy and sustainability* (pp. 234–253). IGI Global Scientific Publishing., DOI: 10.4018/979-8-3693-6567-0.ch012

Marantz, N. J., & Ulibarri, N. (2022). The tensions of transparency in urban and environmental planning. *Journal of Planning Education and Research*, *42*(3), 401–412. DOI: 10.1177/0739456X19827638

Ma, S., Liu, Y., Liu, Y., Wang, J., Fang, Q., & Huang, Y. (2025). Artificial intelligence-enabled predictive energy saving planning of liquid cooling system for data centers. *Advanced Engineering Informatics*, *65*, 103283. Advance online publication. DOI: 10.1016/j.aei.2025.103283

Masanet, E., & Lei, N. (2020). *How much energy do data centers really use*. Aspen Global Change Institute.

Masanet, E., Shehabi, A., Lei, N., Smith, S., & Koomey, J. (2020). Recalibrating global data center energy-use estimates. *Science*, *367*(6481), 984–986. DOI: 10.1126/science.aba3758 PMID: 32108103

Mashaw, J. L. (2018). *Administrative law: The American public law system*. West Academic Publishing.

Mazibuko v. City of Johannesburg, 2009 ZACC 28.

McKee, E. (2025). Procedural justice in water management: A review. *Water (Basel)*, *17*(13), 1987. DOI: 10.3390/w17131987

Medarac, H., Magagna, D., & Hidalgo Gonzalez, I. (2018). Projected fresh water use from the European energy sector. *Disaggregated Fresh Water Withdrawal and Consumption in the EU up to, 2050*.

Meijer, G. I. (2010). Cooling Energy-Hungry Data Centers. *Science*, *328*(5976), 318–319. Advance online publication. DOI: 10.1126/science.1182769 PMID: 20395502

Mekonnen, M. M., & Hoekstra, A. Y. (2016). Four billion people facing severe water scarcity. *Science Advances*, *2*(2), e1500323. DOI: 10.1126/sciadv.1500323 PMID: 26933676

Mitra, B., Datta, S., Kincic, S., Samaan, N., & Somani, A. (2023). Gaps in representations of hydropower generation in steady-state and dynamic models. arXiv. https://arxiv.org/abs/2311.03277

Mittelstadt, B. (2019). Principles alone cannot guarantee ethical AI. *Nature Machine Intelligence*, *1*(11), 501–507. DOI: 10.1038/s42256-019-0114-4

Mytton, D. (2020). Hiding greenhouse gas emissions in the cloud. *Nature Climate Change*, *10*(8), 701. DOI: 10.1038/s41558-020-0837-6

Mytton, D. (2021). Data centre water consumption. npj. *NPJ Clean Water*, *4*(1), 1–8. DOI: 10.1038/s41545-021-00101-w

National Audubon Society v. Superior Court, 33 Cal. 3d 419 (1983).

Naug, A., Sarkar, S., Tuli, S., & Buyya, R. (2025). LC-Opt: Benchmarking reinforcement learning and agentic AI for end-to-end liquid cooling optimization in data centers. arXiv. https://arxiv.org/abs/2511.00116

Oberascher, M., Rauch, W., & Sitzenfrei, R. (2022). Towards a smart water city: A comprehensive review of applications, data requirements, and communication technologies for integrated management. *Sustainable Cities and Society*, *76*, 103442. DOI: 10.1016/j.scs.2021.103442

Obringer, R., Nateghi, R., Knee, J., Madani, K., & Kumar, R. (2023). Contemporary climate analogs project north-south polarization of urban water-energy nexus across US cities under warming climate. arXiv. https://arxiv.org/abs/2306.17050

Obunga, P. O., Rwanga, S. S., Dinka, M. O., & Otieno, B. O. (2025). Review of the emerging technologies in the water sector with a focus on the deployment of Internet of Things solutions. npj. *NPJ Clean Water*, *8*(1), 53. Advance online publication. DOI: 10.1038/s41545-025-00487-x

Organization, W. H., Programme, U. U. N. E., & Health, W. O. f. A. (2022). *One health joint plan of action (2022–2026): working together for the health of humans, animals, plants and the environment*. World Health Organization.

Panteleaki, A. M., Paramanayakam, V., Pentsos, V., Karatzas, A., Tragoudas, S., & Anagnostopoulos, I. (2025). A vertical approach to designing and managing sustainable heterogeneous edge data centers. arXiv. https://arxiv.org/abs/2506.01712 DOI: 10.1109/ISVLSI65124.2025.11130313

Patterson, D., Gonzalez, J., Le, Q., Liang, C., Munguia, L.-M., Rothchild, D., So, D., Texier, M., & Dean, J. (2021). Carbon emissions and large neural network training. arXiv. https://arxiv.org/abs/2104.10350

Pinheiro, A. B., do Prado, N. B., & de Moraes, G. H. S. M. (2024). What drives corporate water disclosure? The role of board composition in Brazil and India. *Sustainable Environment*, *10*(1), 2426836. DOI: 10.1080/27658511.2024.2426836

Privette, A. P., Barros, A., & Cai, X. (2026). Data Centers Water Footprint: The Need for More Transparency. *AGU Advances*, *7*(2), e2025AV002140. DOI: 10.1029/2025AV002140

Qi, S., Milojicic, D., Bash, C., & Pasricha, S. (2023). SHIELD: Sustainable hybrid evolutionary learning framework for data center optimization. arXiv. https://doi .org//arXiv.2308.13086DOI: 10.48550

Qi, S., Milojicic, D., Bash, C., & Pasricha, S. (2023a). MOSAIC: A multi-objective optimization framework for sustainable data center management. arXiv. https://arxiv .org/abs/2311.08583

Qi, S., Milojicic, D., Bash, C., & Pasricha, S. (2023b). SHIELD: Sustainable hybrid evolutionary learning framework for carbon, wastewater, and energy-aware data center management. arXiv. https://arxiv.org/abs/2308.13086 DOI: 10.1145/3634769.3634810

Quach, R., Wang, Y., Jahanshahi, A., Wong, D., & Kim, H. (2025). ECLIP: Energy-efficient and practical co-location of ML inference on spatially partitioned GPUs. arXiv. https://arxiv.org/abs/2506.12598 DOI: 10.1109/ISLPED65674.2025.11261793

Ren, S., ... (2023). The water footprint of artificial intelligence. *arXiv preprint.* https://arxiv.org/abs/2304.03271

Rianisa, A., Anggraini, D. S., & Sekti, B. A. (2026). Role of AI in Consumer Decision-Making Processes. In *Modern Consumer Behavior at the Intersection of AI and Social Media* (pp. 123–151). DOI: 10.4018/979-8-3373-2747-1.ch005

Saheb, T., & Dehghani, M. (2021). Artificial intelligence for sustainable energy: A contextual topic modeling and content analysis. arXiv. https://arxiv.org/abs/2110.00828

Salem, A. K., Taha, A. F., & Abokifa, A. A. (2024). Graph neural networks-based dynamic water quality state estimation in water distribution networks. *Engineering Applications of Artificial Intelligence, 137*, 109426. DOI: 10.1016/j.engappai.2024.109426

Samanta, A., & Jiang, Y., [add remaining authors], et al. (2025). Water footprint of datacenter applications: Methodological implications of manufacturing, operational, and decommissioning phases. In *Proceedings of the ACM Symposium on Cloud Computing*. (pp. 102–110). ACM. DOI: 10.1145/3772052.3772216

Sarkar, S., Tuli, S., & Buyya, R. (2025). Hierarchical multi-agent framework for carbon-efficient liquid-cooled data center clusters. arXiv. https://arxiv.org/abs/2502.08337

Sarker, I. H. (2021). Machine learning: Algorithms, real-world applications and research directions. *SN Computer Science, 2*(3), 160. DOI: 10.1007/s42979-021-00592-x PMID: 33778771

Saxena, D., & Singh, A. K. (2022). A proactive autoscaling and energy-efficient VM allocation framework using online multi-resource neural network for cloud data center. arXiv. https://arxiv.org/abs/2212.01896

Sax, J. L. (1970). The public trust doctrine in natural resource law: Effective judicial intervention. *Michigan Law Review, 68*(3), 471–566. DOI: 10.2307/1287556

Schlosberg, D. (2013). Theorising environmental justice: The expanding sphere of a discourse. *Environmental Politics, 22*(1), 37–55. DOI: 10.1080/09644016.2013.755387

Schmitz, R., Flachsbarth, F., Plaga, L. S., Braun, M., & Härtel, P. (2025). Energy security and resilience: Reviewing concepts and advancing planning perspectives for transforming integrated energy systems. arXiv. https://arxiv.org/abs/2504.18396

Sekti, B. A. (2025). AI Technology Management Concepts and Strategies. In *AI Technology Management and Value Creation* (pp. 1–32). DOI: 10.4018/979-8-3373-2497-5.ch001

Sekti, B. A. (2026). Cybersecurity Challenges in Smart City Development. In *Securing Smart Cities Through Modern Cryptography Technologies* (pp. 75–102). IGI Global Scientific Publishing., DOI: 10.4018/979-8-3373-3326-7.ch003

Shah, A., Fauzi, S., Gining, R. A., Razak, T. R., Jamaluddin, M. N. F., & Maskat, R. (2021). IoT-based monitoring systems for smart cities. *Indonesian Journal of Electrical Engineering and Computer Science, 21*(1), 450–456. DOI: 10.11591/ijeecs.v21.i1.pp450-456

Shahi, P., Heydari, A., Modi, H., Chinthaparthy, L. S. R., Barigala, A., Eslami, B., Tradat, M., Chowdhury, U., Agonafer, D., & Rodriguez, J. (2025). Liquid to Liquid In-Row Coolant Distribution Unit Characterization for High Power Data Center. *Journal of Electronic Packaging, 147*(4), 041109. Advance online publication. DOI: 10.1115/1.4069348

Shahrad, A., Stålhammar, S., & Gulsrud, N. M. (2025). Procedural justice for whom? A case study of a densification project. *Local Environment, 30*(2), 546–566. DOI: 10.1080/13549839.2024.2447757

Shehabi, A., Smith, S. J., Masanet, E., & Koomey, J. (2018). Data center growth in the United States: Decoupling the demand for services from electricity use. *Environmental Research Letters, 13*(12), 124030. DOI: 10.1088/1748-9326/aaec9c

Shi, W., Ma, X., Min, Y., & Yang, H. (2024). Feasibility Analysis of Indirect Evaporative Cooling System Assisted by Liquid Desiccant for Data Centers in Hot-Humid Regions. *Sustainability (Basel), 16*(5), 2011. Advance online publication. DOI: 10.3390/su16052011

Shrimpton, E. A., Hunt, D. V. L., & Rogers, C. D. F. (2021). Justice in (English) water infrastructure: A systematic review. *Sustainability (Basel), 13*(6), 3363. DOI: 10.3390/su13063363

Silander, D. (2019). The European Commission and Europe 2020: Smart, sustainable and inclusive growth. In *Smart, sustainable and inclusive growth* (pp. 2-35). Edward Elgar Publishing.

Silva-Rodriguez, J., & Li, X. (2023). Decentralized micro water-energy co-optimization for small communities. arXiv. https://arxiv.org/abs/2310.01681

Singh, A., Patel, N. P., Ehtesham, A., Kumar, S., & Khoei, T. T. (2024). A survey of sustainability in large language models: Applications, economics, and challenges. arXiv. https://arxiv.org/abs/2412.04782

Stacciarini, J. H. S., & Gonçalves, R. J. A. F. (2025). Data Centers, Critical Minerals, Energy, and Geopolitics: The Foundations of Artificial Intelligence. *Sociedade & Natureza*, *37*(1), e77215. DOI: 10.14393/SN-v37-2025-77215

Stillwell, A. S., & Webber, M. E. (2014). Geographic, technologic, and economic analysis of using reclaimed water for thermoelectric power plant cooling. *Environmental Science & Technology*, *48*(19), 11160–11168. DOI: 10.1021/es405820j PMID: 24625241

Strubell, E., Ganesh, A., & McCallum, A. (2019). Energy and policy considerations for deep learning in NLP. *Proceedings of the 57th Annual Meeting of the ACL*. DOI: 10.18653/v1/P19-1355

Subhash Kumar v. State of Bihar, (1991) 1 SCC 598.

Suresh Babu, C. V. Sathiya Narayana S. (2024). Navigating Uncharted Waters: Overcoming Challenges in Integrating Machine Learning and Deep Learning for Enhanced Clinical Risk. In H. Liu, R. Tripathy, & P. Bhattacharya (Eds.), Clinical Practice and Unmet Challenges in AI-Enhanced Healthcare Systems (pp. 289-310). IGI Global. DOI: 10.4018/979-8-3693-2703-6.ch015

Suresh Babu, C. V. Subha Sree, Prahadeesh, V., Adams, S. R., & Prem Kumar, S. (2024). IoT-Driven Water Management Solutions for Sustainable Agriculture in the Age of Autonomy. In R. Rajappan, N. Gowri Ganesh, J. Daniel, A. Ahmad, & R. Santhosh (Eds.), The Convergence of Self-Sustaining Systems With AI and IoT (pp. 242-260). IGI Global. DOI: 10.4018/979-8-3693-1702-0.ch013

Suresh Babu, C. V., Yadav, S., Ganesh, S., Niraianban, A. R., & Barathi, A. (2025). Logistic Regression-Based Injury Prediction and Prevention in Sports Analytics: An AI Approach. In Choudhury, T., Arya, P., Kotecha, K., Sharma, A., & Um, J. (Eds.), *AI and Machine Learning Applications in Sports Analytics* (pp. 103–114). IGI Global Scientific Publishing., DOI: 10.4018/979-8-3693-5385-1.ch005

Talati, S., Zhai, H., Kyle, G. P., Morgan, M. G., Patel, P., & Liu, L. (2016). Consumptive water use from electricity generation in the Southwest under alternative climate, technology, and policy futures. *Environmental Science & Technology*, *50*(16), 9019–9027. DOI: 10.1021/acs.est.6b01389 PMID: 27768843

The Green Grid. (2011). Water usage effectiveness (WUE): A Green Grid data center sustainability metric (White Paper #35). https://www.thegreengrid.org/en/resources/library-and-tools/238-Water-Usage-Effectiveness-%28WUE%29%3A-A-Green-Grid-Data-Center-Sustainability-Metric

Tuli, S., Ilager, S., Ramamohanarao, K., & Buyya, R. (2021). HUNTER: AI-based holistic resource management for sustainable cloud computing. arXiv. https://arxiv.org/abs/2110.05529

UNESCO World Water Assessment Programme. (2024). The United Nations world water development report 2024: Water for prosperity and peace. UNESCO. https://unesdoc.unesco.org/ark:/48223/pf0000388948

UNESCO. (2023). United Nations world water development report 2023. https://www.unwater.org/publications/un-world-water-development-report-2023

UNESCO. (2023). *United Nations world water development report 2023: Partnerships and cooperation for water*. UNESCO Publishing.

United Nations General Assembly. (2010). The human right to water and sanitation (Vol. A). United Nations., RES/64/292.

Utami, A. S., & Sekti, B. A. (2025). AI in Critical Sectors: Healthcare, Education, Finance, and Security - Ethical Challenges in Generative AI. In *Navigating Social Trust in the AI Era* (pp. 155–186). DOI: 10.4018/979-8-3373-7011-8.ch007

Van der Veen, R., & Smith, A. (2022). Cooling the algorithm: The legal and ecological footprint of hyperscale data centers. *Journal of Digital Sustainability*, *5*(2), 88–104.

Vaughn, M. (2021). ASHRAE Research Report: 2020-2021. *ASHRAE Journal*, *63*(10), 73–87.

Vellore Citizens Welfare Forum v. Union of India, (1996) 5 SCC 647.

Verdecchia, R., Sallou, J., & Cruz, L. (2023). A systematic review of green AI. *Wiley Interdisciplinary Reviews. Data Mining and Knowledge Discovery*, *13*(4), e1507. DOI: 10.1002/widm.1507

Vinuesa, R., Azizpour, H., Leite, I., Balaam, M., Dignum, V., Domisch, S., Felländer, A., Langhans, S. D., Tegmark, M., & Fuso Nerini, F. (2020). The role of artificial intelligence in achieving the Sustainable Development Goals. *Nature Communications*, *11*(1), 233–240. DOI: 10.1038/s41467-019-14108-y PMID: 31932590

Waddenvereniging & Vogelbeschermingsvereniging v. Staatssecretaris van Landbouw, Case C-127/02 (2004).

Wahyuningrum, I. F. S., Budihardjo, M. A., Rustianto, A., & Handayani, N. U. (2023). Determinants of corporate water disclosure in Indonesia. *Sustainability (Basel), 15*(14), 11107. DOI: 10.3390/su151411107

Wang, J., Ulibarri, N., Scott, T. A., & Davis, S. J. (2023). Environmental justice, infrastructure provisioning, and environmental impact assessment: Evidence from the California Environmental Quality Act. *Environmental Science & Policy, 146*, 66–75. DOI: 10.1016/j.envsci.2023.05.003

Wang, Q., Huang, N., Chen, Z., Chen, X., Cai, H., & Wu, Y. (2023). Environmental data and facts in the semiconductor manufacturing industry: An unexpected high water and energy consumption situation. *Water Cycle, 4*, 47–54. DOI: 10.1016/j.watcyc.2023.01.004

Wan, J., Duan, Y., Gui, X., Liu, C., Li, L., & Ma, Z. (2023). SafeCool: Safe and Energy-Efficient Cooling Management in Data Centers with Model-Based Reinforcement Learning. *IEEE Transactions on Emerging Topics in Computational Intelligence, 7*(6), 1621–1635. DOI: 10.1109/TETCI.2023.3234545

Williams, E., Ayres, R., & Heller, M. (2002). The 1.7 kilogram microchip: Energy and material use in semiconductor production. *Environmental Science & Technology, 36*(24), 5504–5510. DOI: 10.1021/es025643o PMID: 12521182

Wood, M. C. (2014). *Nature's trust: Environmental law for a new ecological age*. Cambridge University Press.

World Resources Institute. (2023). Aqueduct Water Risk Atlas. https://www.wri.org/aqueduct

Xiao, T., Nerini, F. F., Matthews, H. D., Tavoni, M., & You, F. (2025). Environmental impact and net-zero pathways for sustainable artificial intelligence servers in the USA. *Nature Sustainability, 8*(12), 1541–1553. DOI: 10.1038/s41893-025-01681-y

Yu, H.-C., Kuo, L., & Ma, B. (2020). The drivers of corporate water disclosure in enhancing information transparency. *Sustainability (Basel), 12*(1), 385. DOI: 10.3390/su12010385

Zhang, G., Zhang, S., Zhang, W., Shen, Z., & Wang, L. (2020). Distributed energy management for multiple data centers with renewable resources and energy storages. *IEEE Transactions on Cloud Computing, 10*(4), 2469–2480. DOI: 10.1109/TCC.2020.3031881

Zhang, L., Tang, Q., & Huang, R. H. (2021). Mind the gap: Is water disclosure a missing component of corporate social responsibility? *The British Accounting Review, 53*(1), 100940. DOI: 10.1016/j.bar.2020.100940

Zhang, Q., Meng, Z., Hong, X., Zhan, Y., Liu, J., Dong, J., Bai, T., Niu, J., & Deen, M. J. (2021). A survey on data center cooling systems: Technology, power consumption modeling and control strategy optimization. *Journal of Systems Architecture*, *119*, 102253. Advance online publication. DOI: 10.1016/j.sysarc.2021.102253

Zhan, X., Zhu, X., Cheng, P., Hu, X., He, Z., Geng, H., Leng, J., Zheng, H., Liu, C., Hong, T., Liang, Y., Liu, Y., & Zhao, F. (2025). Data center cooling system optimization using offline reinforcement learning. arXiv. https://arxiv.org/abs/2501.15085

Zhou, Y., Zhang, Y., Liu, J., & Gao, X. (2020). A review on data center cooling system using artificial intelligence. *Energy and Building*, *227*, 110407. DOI: 10.1016/j.enbuild.2020.110407

About the Contributors

Hitesh Mohapatra received the B.E. degree in Information Technology from Biju Patnaik University of Technology (BPUT), Odisha in 2006, and the MTech. Degree in CSE from Odisha University of Technology and Research (OUTR) University, Odisha in 2009. He received his Ph.D. in Computer Science & Engineering in 2021 from Veer Surendra Sai University of Technology (VSSUT), Burla, India. Dr. Mohapatra has published over 150 research articles in SCI- and Scopus-indexed journals and conferences, with a Scopus H-index of 24. He has authored two books on Software Engineering and C Programming respectively. He has 15 years of teaching experience both in industry and academia. He has served the research community with various capacities like session chair, technical chair, keynote speaker etc. His research interests include wireless sensor network, smart city, smart grid and smart water. Currently he is working as Associate Professor at School of Computer Engineering, KIIT (Deemed to be) University, Bhubaneswar, Odisha, India.

Amiya Kumar Rath received the B.E. degree in computer from Dr. Babasaheb Ambedkar Marathwada University, Aurangabad, in 1990, the M.B.A. degree in systems management from Shivaji University in 1993, the MTech. degree in computer science from Utkal University in 2001, and the Ph.D. degree in computer science from Utkal University, in 2005, with a focus on embedded systems. He is currently a Professor with the Department of Computer Science and Engineering, Veer Surendra Sai University of Technology, Burla, India. He has contributed over 80 research level papers to many national and international journals and conferences. He has published seven books by reputed publishers. His research interests include embedded systems, ad hoc networks, sensor network, power minimization, evolutionary computation, and data mining. He has been deputed as an adviser to National Assessment and Accreditation Council (NAAC), Bangalore, India from

2019 to 2023. Currently he is working Vice Chancellor for Biju Patnaik University of Technology (BPUT), Rourkela, Odisha since 2023 up to 2028.

Fadi Al-Turjman received the Ph.D. degree in computer science from Queen's University, Canada, in 2011. He serves as the "advisor to the chairman of the board of trustees on AI and informatics" with Near East University (NEU), Nicosia, where he is the Founding Dean of the AI and Informatics Faculty. He is the Head of the Software Engineering Department and the Founding Director of the AI and Robotics Institute and the International Research Center for AI and IoT, NEU. His publication history spans over 650 SCI/E publications, in addition to numerous keynotes and plenary talks at flagship venues. He has authored and edited over 100 books on AI, cognition, security, and the deployment of wireless sensor networks in smart IoT environments, which have been published by reputable publishers, such as IEEE, Taylor & Francis, Elsevier, IET, and Springer. He is a leading authority in the areas of smart and intelligent IoT systems, wireless and mobile networks' architectures, protocols, deployments, and performance evaluation in Artificial Intelligence of Things. He has been awarded the Lifetime Golden-Award of Dr. Suat Gunsel from Near East University in 2022, in addition to several other recognitions and best research awards at key international venues. He is leading several international conferences and workshops in prominent AI and IoT societies. He currently serves as a book series editor with distinguished publishers and as the lead guest or an associate editor for several top-tier journals, including the IEEE Communications Surveys and Tutorials (IF 23.9) and Sustainable Cities and Society (Elsevier, IF 10.8).

Manjur Kolhar received the Ph.D. degree in heterogeneous IP communication systems from the National Advanced IPv6 Centre of Excellence, Universiti Sains Malaysia. He is currently an Associate Professor of computer science and information with Prince Sattam Bin Abdulaziz University, Saudi Arabia. His research interests include computer science, computer networks, cloud computing, machine learning, and resource management.

***

Pratibha Agrawal is Assistant Professor in the Department of Commerce and Management at Amity University Jharkhand. Her academic interests lie in interdisciplinary research spanning social sciences, gender studies, cultural studies, indigenous knowledge systems, and community engagement. She has actively participated in academic seminars, workshops, and research initiatives focusing on contemporary social issues, education, and sustainable development. She has been involved in organizing academic events and interdisciplinary collaborations that promote dialogue across diverse fields of inquiry. Her work engages with questions of identity, culture, social transformation, and local knowledge systems within changing socio-cultural contexts.

Mohammad Al-Khaldy is an Assistant professor of Business Intelligence and Data Analytics at University of Petra, Jordan. He completed his PhD in Atrificial Intelligence and Data Science for the University of Hull -United Kingdom - 20017. Alkhaldy's research interests include data analytics, machine learning, predictive analytics, NLP, and decision support systems. His research has been published extensively in academic journals and conferences. Alkhaldy taught a variety of courses in Artificial intelligence and business intelligence, including Data Mining, Business Analytics, Computer Programming, Intelligent business systems, Data Visualization, and machine learning. Alkhaldy is a member of Jordan Computers Society, and Arab Robotics & AI Association. He also a reviewer for several other academic conferences and journals.

Osman Alhussein is an AI Engineer and Lecturer with 4+ years of experience in software development, data analysis, and web technologies. Skilled in Python, TensorFlow, React, and IoT solutions. Passionate about building scalable AI systems, teaching computer science, and applying machine learning to solve real-world problems. Experienced in academic research on neural networks, deep learning, and intelligent systems. Open to collaborations in AI, IoT, and innovative technology projects.

Suman Paliwal is an Associate Professor in the Faculty of Law with 16 years of teaching experience and 5 years of advocacy practice. She holds an LL.M. in Criminal Law and Labour Laws and a Ph.D. in Law. She has published more than 40 research papers in reputed national and international journals, including Scopus-indexed and UGC CARE-listed journals. She is a Member of the Board of Studies and is actively involved in curriculum design and academic planning. She serves as Convener of the Women Development Cell and Presiding Officer of the Internal Complaints Committee, contributing significantly to gender sensitization and institutional governance. Under her supervision, five Ph.D. degrees have been awarded. She has received the Academic Excellence Award from the Bar Council of Rajasthan in recognition of her contribution to legal education and research. Her areas of specialization include Criminal Law, Labour Laws, Women-Centric Laws, Human Rights, and Legal Education.

Reema Bhattacharya, Associate professor in Law at Jagannath University, NCR, she has completed her Ph.D. from Amity Law school, Amity University,

Noida. Dr. Reema is a visiting lecturer for academic session at University of Gdansk, Poland and she is an Awardee of 4th Central Asian Legal Research Fellowship at Tashkent State University of Law, Uzbekistan. She has completed her LLM from Department of Law, Kurukshetra University, Haryana, and prior to that LL. B and BA(H) in Political Science from University of Delhi and she has also completed Master's (M.A) in Public Administration from Jesus and Merry College. Dr. Reema has achieved several recognitions like Won Best Presentation Award at 3rd International Conference, held at Galgotia's University, Noida, Best presentation Award at 3rd World Conference on Human Rights and Intellectual Property Rights held at ILI, New Delhi, Best presentation Award at International Conference organized by Knowledge steez in collaboration with University Pendidikan Sultan Idris and also secured 1st Position in Best Research in International Conference of Forensic Sciences 2025 organized by Amity Institute of Forensic Sciences, Noida. She has also won Young Academic Excellence Award, 2019 at National Award and Symposium, held at PG College, Allahabad. Dr. Reema has also appeared as a Panel Judge at various International & National Moot Court Competitions, International Arbitration and Mediation Competition organized by Jamia Hamdard, DME Law College, Bharti Vidyapeeth University, Delhi, GD Goenka University, and SRMU, etc. She has also completed 12 FDPs Organized by Malviya Mission Teacher Training Program under UGC and various Universities in collaboration with various government bodies. Dr. Reema has represented India as a Delegate at the events like; 5th International Youth Forum at United Nations Conference Centre, Bangkok, Thailand and also at All India Forensic Science Summit, 2025 under Ministry of External Affairs held at Vigyan Bhawan, New Delhi to represent her research topic on Forensic and Terrorism Trials. She has been invited at IP India National Award and Conference, held at Bharat Mandapam as Invitee. Dr. Reema has provided training on gender sensitization and POSH Act 2013 as a speaker at the workshop at Indian Oil Corporation Limited, Regional Division offices, New Delhi and to the employees of Delhi Transco. She is tremendously engaged with sports in her lifetime and participated in National Volleyball tournaments and has secured 1st position & Gold Medal in All India Volley Tournament and 3rd position in Delhi University's Annual Sports Meet and many more. She has published more than 10 Research articles in various SCOPUS, Web of Science, UGC – CARE & Peer Reviewed Journals. She has presented 36 Research Papers in various International & National Conferences and Seminars and also contributed in 7 book chapters.

Youcef Gheraibia is a Senior Lecturer in Computer Science at De Montfort University, Leicester. He holds a Ph.D. in Computer Engineering from the University of Annaba, Algeria (2016), and has previously held positions at York St John University and the University of Hull. His research interests include machine learning, data science, optimization, probabilistic risk and safety analysis, and autonomous system safety, with over 40 peer-reviewed publications to his name. In 2016, he was honored with the National Innovation Award from the Algerian Government.

Dhanvanti Godha is an Assistant Professor at Jagannath University, Jaipur. She has published over 6 research papers and several NPTEL courses . She also has Industrial related experience in Civil Engineering with software skills related to construction oriented work.

Ala Hamarsheh is an associate professor at the Faculty of Engineering and Information Technology of the Arab American University of Jenin. Dr. Hamarsheh has obtained a PhD in engineering sciences from Vrije Universiteit Brussel (VUB)/ Brussels-Belgium in 2012. He graduated in computer science at the Faculty of Science, Birzeit University, Palestine, in 2000. He obtained an MSc degree in computer science at the Kind Abdullah II School for IT, The University of Jordan, Jordan, in 2003. Dr. Hamarsheh has published numerous papers in international refereed journals and conferences.

Balamurugan K. is a highly regarded academic and researcher with over 29 years of expertise in Information Technology and Computer Science. Currently, he is an Associate Professor in the Department of Information Technology at K.S.R. College of Engineering (Autonomous) in Tiruchengode, Namakkal Dt, Tamil Nadu. He completed his B.E. in Electronics and Communication Engineering from Kumaraguru College of Technology in 1995, M.E. in Multimedia Technology from the College of Engineering, Guindy, Anna University, in 2003, and earned his Ph.D. in Information and Communication Engineering from Anna University in 2018.

Ganesh Kumar K working as a Professor in the department of Information Technology, Velalar College of Engineering and Technology, Thindal, Erode, Tamilnadu, India with 16 years of teaching experience. He has published 15 articles in various national and international level journals. He is an active review member in various reputed journal. He has guided more than 25 undergraduate and 15 post graduate level students projects.

Keerthana N V is an Assistant Professor with 8 Years of experience in teaching and 1-year experience in research (IUCAA LIGO Data Centre, Pune). Her area of interest is Network Security, Blockchain, and Machine Learning Algorithms. She is pursuing PhD at Anna University in Blockchain Security. She has work experience in high-performance computing (HTCondor) for the LIGO INDIA Project.

Naveen Nandal, Associate Professor at DIHE, Rohini, Delhi. Dr. Naveen Nandal is a distinguished academic and research professional with over 12 years of experience in teaching, research, and analytics. Currently an Associate Professor at DIHE, Delhi, she has published over 80 research papers, authored six books, and holds several patents. She has contributed extensively to Scopus-indexed, ABDC, WOS, and peer-reviewed journals. She has served as Analytics Area Chair and Research & Publication Head, actively involved in curriculum design, faculty development, and research guidance. Dr. Nandal is a reviewer for Sage and Emerald publications, a founding member of the Research India Foundation, and has extensive experience delivering impactful training and mentoring in marketing, human resource management, and data analytics domains.

Periyasamy S is working as an Associate Professor in the Department of Computer Science and Engineering, Selvam College of Technology, Namakkal with 15 years of teaching experience.

Sujan S Pursuing his second year B.Tech Information Technology in Velalar College of Engineering and Technology, Thindal, Erode.

Dini Febryana Sari is an Information Technology professional with a profound research interest in Artificial Intelligence (AI) and the Internet of Things (IoT). With a strong academic background and professional experience in digital technology environments, she actively explores the integration of AI and IoT to drive innovative solutions for automation, operational efficiency, and digital transformation. Her research portfolio includes work in intelligent analytics, sensor data processing, and the development of connected system architectures. She is committed to advancing the digital innovation industry through rigorous research, focusing on how emerging technologies can be leveraged to create sustainable and impactful technological progress.

Binastya Anggara Sekti has been a lecturer at the Faculty of Computer Science, Esa Unggul University, Jakarta, since 2016. He is actively involved in the Artificial Intelligence and Information Security Study Group. He holds a Bachelor's degree in Informatics Engineering from the Bandung Institute of Technology and a Master's degree in Information Systems Management from Trisakti University, Jakarta. Prior to entering academia, he accumulated over 25 years of professional experience as an IT practitioner in various companies across Jakarta. His work focused on developing data centers for internet service providers and delivering IT solutions for the manufacturing and oil & gas sectors. His research interests include artificial intelligence, cybersecurity, and the development of IT infrastructure.

Arti Sharma is an accomplished academician and legal expert, currently serving as the Head of the Department of Law at Jagannath University, Bahadurgarh, Delhi NCR. Since assuming this role in 2022, she has been instrumental in fostering a culture of excellence and innovation in legal education and research. Dr. Sharma's academic journey began with an integrated B.A. LL.B. program at Maharshi Dayanand University, Rohtak, Haryana. Driven by her passion for law, she pursued an LL.M. in Criminal Law at the same institution. She further solidified her academic credentials by earning a Ph.D. in Law from Chaudhary Devi Lal University, Sirsa, where her research delved into significant legal issues, showcasing her analytical depth and dedication to the field. Academic and Administrative Contributions Throughout her career, Dr. Sharma has consistently taken on roles that extend beyond teaching. At Jagannath University, she is an active member of the Internal Quality Assurance Cell (IQAC) and the Academic Council, playing a pivotal role in enhancing academic standards and governance. Additionally, she serves on the Internal Complaint Committee (ICC), where her contributions have been vital in ensuring a safe and inclusive campus environment. Expertise in Workplace Harassment and Gender Equality Dr. Sharma's expertise in addressing workplace harassment and promoting gender equality has made her a sought-after speaker and trainer. She has delivered numerous lectures on the Sexual Harassment of Women at Workplace (Prevention, Prohibition & Redressal) Act, 2013 (POSH Act) across various institutions. Dr. Sharma has conducted Gender Sensitization and POSH Act training sessions at: • Indian Oil Corporation Limited, Delhi • Indian Oil Bhavan, Yusuf Sarai, Delhi • Various government schools in Jhajjar, Haryana These sessions are designed to equip participants with the knowledge and tools needed to foster respectful and equitable work environments. Her practical insights and ability to communicate complex legal frameworks make her training programs impactful and highly relevant. Research and Scholarly Contributions Dr. Sharma is a prolific researcher with extensive publications in UGC-CARE listed and peer-reviewed journals. She has also contributed to book chapters and presented papers at national and international conferences. Her scholarly work reflects a deep engagement with critical legal issues and highlights her ability to address diverse academic audiences with clarity and authority. Mentorship and Student Development Mentorship is a cornerstone of Dr. Sharma's professional ethos. She has successfully guided five Ph.D. scholars to completion and has supervised numerous undergraduate and postgraduate dissertations. She has also served as a judge in a Moot Court Competition organized by the Centre for Professional and Allied Studies, Maharshi Dayanand University, Gurugram, demonstrating her commitment to supporting legal education through experiential learning platforms. Her mentorship empowers students to engage in rigorous academic research and develop a nuanced understanding of legal principles, preparing them for impactful careers in the legal field. Continuous Professional Development Dr. Sharma is committed to lifelong learning and professional growth. She actively participates in workshops, seminars, conferences, and faculty development programs. These engagements ensure that she remains at the forefront of legal education and practice, bringing fresh perspectives and contemporary knowledge to her teaching and research endeavours. Vision and Leadership As the Head of the Department of Law, Dr. Sharma has played a pivotal role in enhancing the quality of legal education at Jagannath University. Under her leadership, the department has flourished as a hub of academic excellence, fostering critical thinking, innovation, and ethical practices among students and faculty alike.

C.V. Suresh Babu is a pioneer in content development. A true entrepreneur, he founded Anniyappa Publications, a company that is highly active in publishing books related to Computer Science and Management. Dr. C.V. Suresh Babu has also ventured into SB Institute, a center for knowledge transfer. He holds a Ph.D. in Engineering Education from the National Institute of Technical Teachers Training & Research in Chennai, along with seven master's degrees in various disciplines such as Engineering, Computer Applications, Management, Commerce, Economics, Psychology, Law, and Education. Additionally, he has UGC-NET/SET qualifications in the fields of Computer Science, Management, Commerce, and Education. Currently, Dr. C.V. Suresh Babu is a Professor in the Department of Information Technology at the School of Computing Science, Hindustan Institute of Technology and Science (Hindustan University) in Padur, Chennai, Tamil Nadu, India.

Sangamithra T is working as an Assistant Professor in the Department of EEE, Velalar College of Engineering and Technology, Thindal, Erode with 6 years of experience in teaching.

Akanksha Yadav is Assistant Professor of Sociology at Amity University Jharkhand. She completed her Ph.D. from the Department of Humanistic Studies, IIT (BHU), Varanasi, where her research explored the lived religious experiences, ritual practices, and folk traditions of Cha ha Pūjā through ethnographic fieldwork conducted across Bihar, Uttar Pradesh, and Jharkhand. Her research interests include gender studies, religion, folklore, environmental humanities, media and advertising studies, menstruation studies, and indigenous knowledge systems. She has published in national and international journals and has presented her work at several academic conferences, including ECSAS. She is also associated with editorial and academic bodies in gender studies.

Index

A

B

C

D

E

F

I

L

M

P

R

S

T

W

www.ingramcontent.com/pod-product-compliance
Lightning Source LLC
LaVergne TN
LVHW081521300926
849161LV00006B/1260
9798260027615